TJ   Asada, H.
211   (Haruhiko)
.A78
1987   Direct-drive robots

$30.00

DATE			

Direct-Drive Robots
Theory and Practice

Direct-Drive Robots
Theory and Practice

Haruhiko Asada

Kamal Youcef-Toumi

The MIT Press
Cambridge, Massachusetts
London, England

© 1987 Massachusetts Institute of Technology

This book was printed and bound in the United States of America.

Library of Congress Cataloging-in-Publication Data

Asada, H. (Haruhiko)
 Direct-drive robots.

 Bibliography: p.
 Includes index.
 1. Robotics. 2. Manipulators (Mechanism)
I. Youcef-Toumi, Kamal.   II. Title.
TJ211.A78 1987   629.8'92   87-3174
ISBN 0-262-01088-7

To Our Parents and Families

Contents

II. ARM DESIGN THEORY

List of Photos

List of Figures

List of Tables

Preface

The field of control engineering has provided valuable theoretical contributions to advancements in Robotics. Aspects of control engineering which have been utilized include: linear systems theory, multivariable control, adaptive control and nonlinear control. The need for further development and research in these and related areas is apparent when new applications and more stringent performance criteria are encountered. In fact, some fundamental and somewhat rudimentary issues have not yet been resolved.

The simplest and most common compensation technique for industrial robots is the PID type controller. In general, this controller is designed with the assumption of independent manipulator joint dynamics, thus its performance is limited because manipulators are n degree-of-freedom coupled nonlinear dynamic systems. The PID controllers, however, are generally designed and easily implemented for single-input/single-output plants. Although the overall system's performance using these controllers may not be acceptable, this method has been proven to guarantee global stability of robot manipulators and can lead to acceptable performance in some cases. To improve the performance, feedforward compensation has been introduced to effectively correct for tracking errors. This approach works reasonably well when the plant parameters have been identified correctly. In order to maintain acceptable responses in the face of changing parameters or loads, adaptive control schemes have been used to re-tune the controller gains continuously. Also, several other schemes have been proposed for the control of robot manipulators.

The performance of a system, however, does not depend on the control system design alone. An appropriate match between system hardware design and controller design is a key issue, particularly when high performance is required. Thus, in the design of machines for advanced applications, fundamental physical understanding should reflect all aspects of the control system including those relevant to mechanical design. The unification of theoretical work with mechanical design is necessary to improve system hardware and ultimately to permit "exact" execution of the control actions. These considerations are of great importance in the development of precision positioning robots for submicron assembly operations, and high speed high accuracy manipulators for trajectory control. For example, in certain laser cutting applications the end effector speed, acceleration and tracking tolerance are on the order of 3 m/s, 3 to 5 G and 0.05 mm respectively. Thus, the hardware for such applications must be properly engineered to meet these severe control specifications.

The gap between mechanical design and control is gradually closing. Advancements in control engineering have contributed strongly to the merging of these fundamental issues, and the development of a unified approach to robot design and control. The goal of this book is to present such a unified approach to design and control in the development of high performance robot manipulators -- this text is intended to provide not only theoretical fundamentals needed for analysis and synthesis but also practical hardware implementations in direct-drive robot technology.

This book is organized into four parts: Direct-Drive Technologies, Arm Design Theory, Development of M.I.T. Direct-Drive Manipulators, and Supporting Articles.

Chapter 1 of this book starts with a historical perspective in robot design, then presents the direct-drive concept. Description of several direct-drive robots is given as an overview for the state-of-the-art in this technology. This includes a few models that were designed at the M.I.T. Laboratory for Manufacturing and Productivity, Carnegie-Mellon University and industrial companies that have produced products which use this concept.

The robot components; such as motors, drive amplifiers, sensors and arm linkages; are covered in Chapter 2. More importantly, the issues in arm design and control are also discussed in this chapter.

Chapters 3, 4, 5 and 6 deal with arm design theory. They introduce analytical tools for evaluation of the static characteristics of manipulators and new approaches in the design of manipulators with simplified dynamics.

The development of the M.I.T. direct-drive robot for high speed trajectory control is presented in chapters 7 and 8. These chapters cover not only the design of the mechanism but also the control system design.

The last part of this book presents some supporting articles relevant to the design and control of direct-drive manipulators.

We would like to express our sincere thanks to Professor David N. Wormley, Head of the Department of Mechanical Engineering at M.I.T, for his support and encouragement since we began writing this book in the early summer of 1985.

Special thanks go to Professors Steven Dubowsky, Neville Hogan and John Hollerbach of M.I.T for their comments and criticisms on parts of this book while serving as members on Youcef-Toumi's ScD thesis committee. The support of Professor David E. Hardt, Director of the M.I.T Laboratory For Manufacturing and Productivity is greatly appreciated.

We deeply appreciate the intellectual stimulation and financial support of our Industrial sponsors Shin Meiwa Industry Co., LtD and Matsushita Electric Industrial Co., Ltd of Japan. Without their support this work would not have been possible. Our thanks go to Dr. Makio Kaga, Director, and Mr. Hideo Koyama, Chief Engineer, and Mr. Hirotoshi Yamamoto, Design Engineer, all with Shin Meiwa's Development Center. To Mr. Shinya Yamauchi, General Manager, and Mr. Tetsuya Takafumi, Engineer, of Matsushita.

The financial aid provided by the M.I.T Gordon Support is greatly appreciated.

We would like to thank Professor George Chryssolouris for the generous use of his printer and Ms. Teresa Ehling of M.I.T Press for her patience with our delayed drafts of this book and her suggestions regarding the book formats. We thank Mr. Neil Goldfine for proofreading the material presented in this book, and Bob Brodersen, Frank Chiricosta and Wesley Cobb for typing the manuscript.

Part I

DIRECT-DRIVE TECHNOLOGIES

Chapter 1

Introduction

1.1 Historical Perspectives of Robot Design

In the early 1960's, Unimation Inc., in Danbury, Connecticut, made the first successful installation of what we call an industrial robot today. Their robot was a five-axis, hydraulically driven manipulator arm capable of carrying a several-pound payload along a predetermined path. After a careful task analysis, their robots were installed for the loading and unloading of workpieces for dicasting machines. The task involved repetitive, tedious work in a dirty, noisy and dangerous environment. The robot, capable of repeating instructed motions, met the increasing requirement for the replacement of unpleasant manual operations.

The mechanical construction of the first Unimate industrial robot departed from traditional machine design in many ways. The arm was basically a cantilevered beam structure with many degrees of freedom, most of which were revolute joints that formed a closed kinematic chain. This structure allows the robot to perform flexible motions and access a large work space compared with the space occupied by the robot itself. Also, the robot arm can operate in a crowded area as we have seen in most factory environments. Thus, the open kinematic chain meets the high mobility and flexibility requirements in the kinematic structure design. However, this kinematic structure created a number of difficult machine design problems. The positioning accuracy at the endpoint of the long cantilevered arm is considerably low; and the mechanical stiffness of the arm construction is inherently poor as well. Also, a small amount of error at each revolute joint is magnified at the arm's endpoint as the arm length gets longer. As a result, the required accuracy for the actuator driving the revolute joint must be significantly high for the articulated arm. The hydraulic actuators used for the first Unimate robot were a reasonable choice because of the high performance required for the drive system. Hydraulic drives are generally high precision, high power actuators with particularly high torque-to-weight ratio. In consequence, hydraulic robots are capable of accurate motions under severe load conditions.

As applications of industrial robots expanded, different types of robots were developed to meet a variety of task requirements. For light-duty applications, electrically-powered robots became the most prominent robot design. In arc welding, for example, electrical robots are most widely used. Compared with hydraulic drives, electro-mechanical drives are, generally, inexpensive and clean, as well as easy to maintain. Thus, electrical robots continue to replace hydraulic robots in many applications.

Electrical motors generally produce their maximum power at high speed. In other words, the electrical motors exert rather small torques while rotating at high speeds. In consequence, appropriate gearing is necessary for the electrical motors in order for these systems to drive such loads. Robot arms are usually moved at low speeds, less than one revolution per second, while required maximum torques range from a few newton-meters to several hundred newton-meters. A large gear reduction on the order of 1:100 is typically required for standard servo motors.

Design of electro-mechanical drives is complicated by the need for gearing and reducers. The reducer must provide a high gear reduction while maintaining high precision. Particularly important is that the reducer should not introduce backlash or lost motion, which will directly degrade positioning accuracy. Even a small amount of backlash at a proximal joint leads to a significantly large error at the arm tip. Hence, backlash must be completely eliminated at each joint.

Friction is another problem in the design of gearing. As described later, anti-backlash gears and other types of reducers with low backlash are characterized by considerably large friction. In many cases the motor power dissipated at the gearing amounts to over 30% of the total power because of the large friction. The large friction also leads to poor control accuracy, which is a more serious problem for high precision applications. Friction is an unpredictable characteristic, and is difficult to identify, hence difficult to compensate for.

Gearing mechanisms are often major sources of mechanical deflections or compliances. Poor mechanical stiffness not only causes static arm deflections but also limits dynamic responses. The loop gain of the servo system cannot be increased if the higher order delay resulting from the low stiffness makes the system unstable. This severely limits the bandwidth and accuracy of the drive system. Poor stiffness at the gearing may also cause undesirable vibrations.

This vibration problem is a critical issue particularly for high speed manipulation . After arriving at a specified point, the robot cannot proceed to the following step of motion, until the residual vibration caused by the gearing compliance ceases. Thus, the reduced mechanical stiffness is a crucial problem when the tact time of operations is short.

Traditionally, speed and accuracy are main issues in the design of machines. Robots are, however, somewhat different in that the speed and accuracy are not necessarily the primary goal of design. Essential robot characteristics are described by such words as flexibility, dexterity, intelligence, etc. Consequently, much research effort has addressed issues regarding how robots can be more flexible, dexterous and intelligent.

In advanced manipulation tasks, such as assembly of mechanical parts, robots perform the task through mechanical interactions with the environment. The robot is required to accommodate contact forces as well as to control the location of its end effector. Force control and compliance control are then necessary for those tasks which require control of mechanical interactions. Control schemes such as these, have been implemented on traditional robots that were primarily designed as positioning devices. The drive systems of these devices are basically position-controlled, and are not necessarily appropriate for force control . Reducers in electro-mechanical drives create significantly large friction , which provides disturbance torques in the force control system and degrades the control performance. It is also difficult for a hydraulic robot to control delicate interaction forces between the arm tip and the environment. Thus, we need an appropriate hardware tool for delicate force control and advanced manipulation.

1.2 The Direct-Drive Approach

Direct-drive is, basically, an electrical drive in which no gear reducer is used. The rotor of an electrical motor is directly coupled to the load, hence the mechanical gearing is completely eliminated. The direct-drive robot is defined to be a mechanical arm where all or part of the active arm joints are actuated with the direct drive.

Figure 1-1 shows the basic construction of a direct-drive joint from a direct-drive robot. The direct-drive joint consists of a pair of arm links, the motor, and the bearings. The motor is comprised of the stator and the rotor.

The stator is housed in the case connected to one of the arm links, usually a proximal link, and the rotor is directly coupled to the joint shaft, which is connected to the other arm link, usually a distal link. Thus the distal arm link is rotated directly by the torque exerted between the rotor and the stator, hence direct drive.

The problems that the mechanical gearing unavoidably possesses can be solved completely in the direct drive method. Backlash is completely removed and friction is reduced significantly. It may exist only at the bearings supporting the joint shaft. The mechanical construction can be much stiffer than drive mechanisms with gearing, wear of gears is no longer troublesome, and the simple construction is more reliable and is easy to maintain.

These features may result in excellent control performance. The accuracy of positioning can be improved remarkably. Since the simple mechanical construction possesses less uncertainties, eg. low friction and no backlash, the repeatability of positioning can be an order of magnitude better than that of the geared drives. The simplified mechanism will also allow us to identify the system dynamics with little difficulty. The behavior of the drive system can be modeled accurately, and thus the dynamic response is more predictable. These desirable static and dynamic characteristics make it easy and effective to apply advanced control schemes including force control and compliant motion control. Thus the direct-drive robot can be a desirable test bed for advanced manipulation studies.

1.3 The State of the Art of Direct-Drive Robots

The basic concept of a direct-drive robot was first established by H. Asada in 1980. He and his colleague T. Kanade developed the first prototype in 1981 at the Robotics Institute, Carnegie-Mellon University. Photo 1-1 shows their robot, CMU D-D Arm Model I, a six degree-of-freedom manipulator arm consisting of all direct-drive revolute joints. The first joint, the most proximal joint at the ceiling, causes a rotation about the vertical axis, while the second joint rotates the upper arm about a horizontal axis. The third joint rotates the forearm about the center line of the upper arm. The motor driving the third joint is fixed to the upper arm and is located on the other side of the elbow part, so that the motor acts as a counterweight of the elbow part. The gravity load of the second motor is then reduced with this mass balancing procedure. The

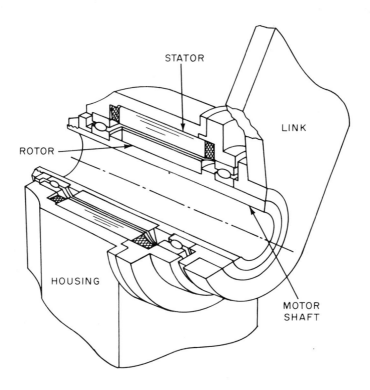

Figure 1-1:   Basic construction of a direct-drive joint

fourth joint is the elbow, causing the bending motion of the forearm, while the fifth joint rotates the forearm about its center line. The motor of the fifth joint is located within the elbow part, so that the load of the fourth motor can be reduced. The last joint is at the wrist part, bending the arm tip about the axis perpendicular to the forearm.

The motors used for the first three joints are DC torque motors using ALNICO magnets at the stators. The maximum torque of the largest motor, that is, the first motor, is over 150 Nm, while the diameter of the stator is 60 cm. The motors of the last three joints are also DC torque motors, but the permanent magnets used are Samarium-Cobalt magnets, which are more than three times stronger than the ALNICO magnet in terms of maximum magnetic energy product .

The first prototype development at the CMU Robotics Institute demonstrated the potential advantages and feasibility of the direct drive approach. The prototype design, however, revealed that the robot of this new type needs more powerful motors with more compact sizes. The motors used for the CMU Arm are not sufficient to drive the large loads, while the size of each motor is so large that the arm construction becomes too bulky and heavy.

At the Massachusetts Institute of Technology, improved direct-drive robots were developed by H. Asada and K. Youcef-Toumi. The motors used for the M.I.T. Arms are brushless DC torque motors with Samarium-Cobalt magnets, which can exert 3 to 10 times larger torque compared with the CMU Arm. For example, the largest motor of one of the M.I.T. arms can produce 660 Nm of output torque, while its stator diameter is only 35cm. In comparison with the largest motor of the CMU Arm, the peak torque is therefore 3 times larger, while the diameter reduces to a half of the previous motor.

Photo 1-2, shows the M.I.T. D-D Arm Model I , a three degree-of-freedom, serial link manipulator arm. The arm was built during the period 1982 to 1983, and is currently installed at M.I.T.'s Artificial Intelligence Laboratory. The robot has a unique kinematic structure, which allows it to eliminate gravity loads at individual motors. The first joint rotates the whole upper body about the vertical axis, hence no gravity load acts upon the first motor. The second joint rotates the forearm about the center line of the upper arm, similar to the third joint of the CMU Arm , but the motion of the upper arm is constrained within a horizontal plane. The third joint is located at the elbow, and causes the bending motion of the forearm. Since the mass of the forearm is balanced with its counterweight, the third motor does not experience any gravity load .

degrees of freedom	6
material	Aluminum
special features	Serial drive

Photo 1-1: The Carnegie-Mellon University Direct-Drive Arm model I

Similarly, the axis of the second joint passes through the centroid of the forearm, hence no gravity load acts on the second motor. Thus all three motors have no gravity load for all arm configurations.

Photo 1-3 shows the M.I.T. D-D Arm Model II, which was developed by the same group at M.I.T. from 1983 through 1984. For this arm, they employed a parallel drive mechanism with a closed-loop kinematic chain. The upper two motors, located on the base frame, drive the two input links of the parallelogram mechanism, which causes a two degree-of-freedom vertical motion at the arm tip. One of the features of this arm construction is that the heavy motor at the elbow joint of the previous model was replaced by the motor mounted on the base frame and that the drive torque is transmitted through the parallelogram mechanism. This remote drive mechanism reduces the arm weight significantly and improves the dynamic performances. Another feature of the Model II D-D Arm is the innovative technique for dynamic mass balancing and decoupling, which are to be discussed in detail in Chapters 4 and 5.

As the velocity of the arm motion increased, the dynamic characteristics of the arm construction have more critical influence upon the control performance. Photo 1-4 shows the M.I.T. D-D Arm Model III with a linkage made out of graphite composite material. Since the material is lightweight and strong, the arm's stiffness as well as the inertia are improved significantly. The natural frequency of the arm construction was then increased to about 70 Hz, whereas the Model II Arm, which was made of aluminum, had a natural frequency of only 14 Hz. With this lightweight, high stiffness arm structure, the M.I.T. D-D Arm Model III has achieved an extremely fast motion: The maximum tip speed is 12 m/s and the maximum acceleration is over 5 G.

Photo 1-5 shows another type of direct-drive arm developed at M.I.T., the M.I.T. D-D Arm model IV. The two motors aligned on a vertical axis drive the horizontal parallelogram mechanism. Since the link motion is constrained within a horizontal plane, no gravity load acts upon the two motors. Robots with this type of kinematic construction are often referred to as SCARA robots , which are widely used for assembly operations, particularly for planar assembly tasks.

The simple construction of the SCARA robot is appropriate for direct-drive robots: small number of degrees of freedom as well as the no- gravity-load construction make it easier to build practical robots. In fact, the first commercialized direct-drive robot employed the same kinematic construction as the SCARA robot. Photo 1-6 shows the AdeptOne direct-drive robot developed

degrees of freedom	3
material	Aluminum
special features	gravity balanced, serial drive

Photo 1-2: The M.I.T. Direct-Drive Arm Model I

Direct-Drive Robots

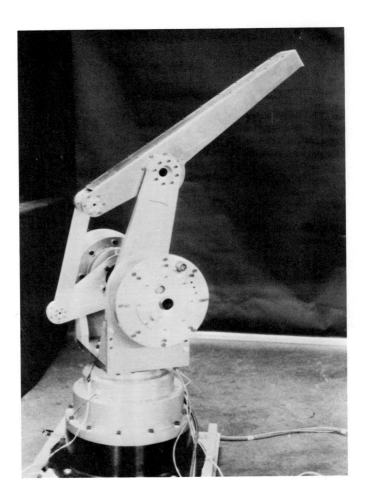

degrees of freedom	3
material	Aluminum
special features	decoupled dynamics, parallel drive with a five-bar-link mechanism

Photo 1-3: The M.I.T. Direct-Drive robot Model II

degrees of freedom	3
material	Graphite Composite
special features	decoupled dynamics, parallel drive with a six-bar-link mechanism
application	laser cutting

Photo 1-4: The M.I.T. Direct-Drive robot Model III

by ADEPT Technology Inc., at Sunnyvale, California, in 1983. The arm has four degrees of freedom. Two motors are located at the base, which produce the horizontal link motion. The other two motors are on the forearm: one drives the lead screw which produces a translational motion along the vertical axis, while the other rotates the gripper about the vertical axis.

To reduce the arm weight, some of the joints are driven remotely by motors located at proximal positions. The second joint, that is the elbow joint, is driven by one of the base motors through a steel band between the motor and the elbow joint. The wrist joint is driven by a motor at the elbow position through another steel band. Though transmission mechanisms are used, no gear reducers are employed in this arm design. In consequence, the features of the direct-drive approach are mostly maintained.

The maximum speed of the arm is 9 m/s, which is defined to be the tip speed when all the joints are at slew speed. The repeatability, that is, the maximum positioning error when reaching the same destination along the same trajectory, is determined to be ± 0.001 inch or ± 0.0254 mm. Both specifications are an order-of-magnitude better than traditional robots with gear reducers.

The motors used for the AdeptOne direct-drive robot are specially designed high torque motors produced by Motornetics Inc. in Santa Rosa, California. As described in detail in the following chapter, the motor is a 3-phase variable reluctance motor having many teeth at the rotor. The motor produces a large torque with small power consumption, when rotating at low speeds.

Photo 1-7 shows another direct-drive robot commercialized by the Matsushita Electric Industrial Co., Ltd., Osaka, Japan. The robot has the same kinematic construction as that of the M.I.T. Arm Model IV, namely the horizontal parallelogram mechanism. The arm length is about 60 cm, a little smaller than the AdeptOne robot. The motors used for the Matsushita robot are brushless torque motors with Samarium-Cobalt magnets . One of the features of the Matsushita robot is that the motor, when incorporated with a high performance drive amplifier , has an excellent linearity, while producing a large torque with a small torque ripple. This allows us to improve the control accuracy not only positioning control but also for the torque control. One can also change the endpoint compliance in a wide range; it can be extremely soft or hard by simply changing the gains of the drive system.

Another feature of the Matsushita robot is its high accuracy. Precision

degrees of freedom	2
material	Aluminum
special features	decoupled and invariant dynamics, parallel drive with a five-bar-link mechanism
application	assembly

Photo 1-5: The M.I.T. Direct-Drive robot Model IV

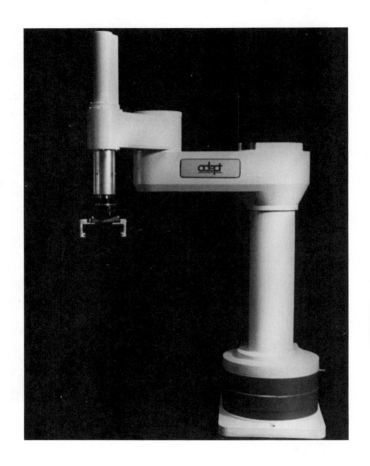

Photo 1-6: AdeptOne direct-drive robot
(Courtesy of Adept Technology, Inc.)

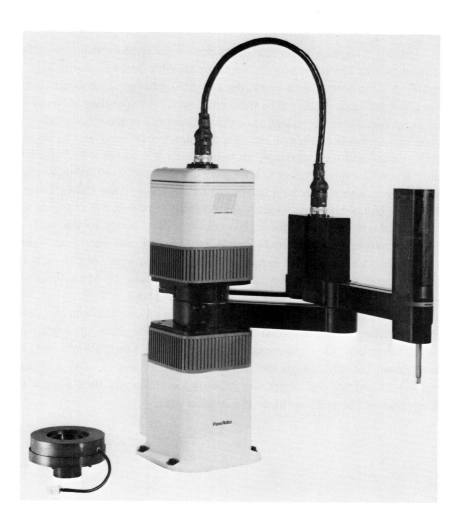

Photo 1-7: The Matsushita Direct-Drive Robot
for high speed assembly, Pana Robo HDD-1.
(Courtesy of Matsushita Industrial Co., Ltd.)

position sensors were specially developed for the direct-drive robots. The sensor is a laser interferometer type encoder, having a high resolution . With this encoder and the motor drive system, the repeatability of the Matsushita robot is less than ± 0.010 mm or ± 0.0004 inch. This will meet increasing needs for high accuracy operations, particulary in precision assemblies of electric parts.

Photo 1-8 shows a larger size direct-drive robot used for the laser cutting of sheet metal. The robot was developed by the Shin Meiwa Industry Ltd., in Japan. A laser beam passes inside the arm links, changes its direction at the arm joints and impinges against a sheet metal part. The speed and accuracy of the metal cutting are dependent upon the performance of the robot used. The direct-drive robot is capable of tracing a complex spacial curve at a high speed while maintaining high accuracy. In laser cutting it is particularly difficult to move around a sharp corner or to trace a circular curve of small radius, because an extremely large acceleration is required for the robot. Shin Meiwa's direct-drive laser cutting robot has a maximum acceleration of over 5 G at the arm tip.

To achieve the extremely high acceleration, the arm links are made of graphite composite, which reduces the arm weight and increases natural frequencies. A special dynamic mass balancing technique was employed in the arm linkage design. These techniques used for both robots by Shin Meiwa and Matsushita will be discussed in detail in Part II of the book.

Thus research and development of direct-drive robots have been conducted both in industry and in academia during the past several years. It ranges from the development of components, such as motors and sensors, to the development of arm design and necessary control theory. The following chapters investigate the component technologies first, and then discuss the design and control problems of the direct-drive arms.

Photo 1-8: The Shin Meiwa Direct-Drive robot for high speed laser cutting applications. (Courtesy of Shin Meiwa Industry Co., Ltd.)

Chapter 2

Drive Systems

2.1 Introduction

Key components in the design and control of a direct-drive robot are the motors and drive amplifiers. An order of magnitude larger torque must be generated in a compact and lightweight body. Fluctuations in torque and speed should be minimized in order to achieve accurate control. There are many other requirements for motors used for the direct-drive robot. In this section we analyze the characteristics of motors and discuss some performance standards, which are critical to the direct-drive robot applications. Motors used in direct-drive robots are:

1. Direct current torque motors ,

2. Brushless D.C. torque motors ,

3. Variable reluctance motors .

We investigate each type of the direct-drive motors and compare their performances.

2.2 Direct Current Torque Motors

2.2.1 Design

Direct current motors are the simplest but most highly efficient motors widely used for robot drives. To apply the D.C. motor to the direct drive robot, the motor design must be modified so that a sufficiently large torque can be obtained. The motors used in the early versions of direct-drive robots, e.g. CMU Direct-Drive Arm , are referred to as D.C. torque motors, or D.C. torquers, which are capable of exerting much larger torques compared with regular D.C. motors . There is, however, no essential difference between regular D.C. motors and the direct-drive D.C. motor in principle; the latter motor is designed in such a way that the output torque, rather than power, be maximized.

As shown in Photo 2-1, a typical D.C. torque motor consists of a stator, a rotor and a brush ring. The stator is made of permanent magnets cemented on a metal ring, while the rotor has windings having many poles on it. The torque is exerted at the air gap between the rotor and the stator. The total area of the air gap A is determined by the product of the length of the rotor, L_r, and the diameter of the air gap cylinder, which is approximately the same as the rotor diameter, D_r,

$$A = L_r D_r \tag{2.1}$$

The output torque is then determined by the air gap area and the rotor diameter,

$$T = k_a A D_r = k_a L_r D^2_r, \tag{2.2}$$

where k_a is a constant determined by the magnetic field at the air gap times the current flowing into the windings . Thus the output torque is proportional to the square of the rotor diameter. Therefore a straightforward way of increasing the torque is to enlarge the rotor diameter rather than increasing the rotor length . This is why the direct-drive torque motor in Photo 2-1 has a large diameter and a short rotor length, which looks like a pancake.

2.2.2 The Torque Constant

The dynamic equation of the D.C. torque motor is the same as that of a regular D.C. motor. The electrical relationship of the armature is given by

$$V = Ri + E + L\frac{di}{dt} \tag{2.3}$$

where V is the voltage applied to the armature, R and L are the resistance and the inductance of the windings, respectively, i is the current, and E is the back emf, which is given by

$$E = K_t \omega \tag{2.4}$$

where K_t is called back e.m.f constant, and ω is the angular velocity of the

Photo 2-1: D.C. torque motor

rotor.

The electrical power converted into mechanical power is given by $P = Ei$, that is, the voltage drop due to the back emf times the armature current. This amount of power is the same as the mechanical power produced by the rotor:

$$Ei = \tau\omega \tag{2.5}$$

Substituting Equ. (2.4) into Equ. (2.5), we obtain the relationship between the torque and the current

$$\tau = K_t i \tag{2.6}$$

where K_t is called torque constant , which is the same number as the back emf constant when SI units are used.

The torque constant is determined by the magnetic field and the geometrical parameters of the air gap. For an ideal torque motor, the torque constant is truly invariant. However in reality, it varies with the rotor position. Since the magnetic field consists of discrete poles and the commutation occurs discretely, the torque constant varies as the rotor rotates. This causes fluctuations in the output torque. The amount of fluctuation is represented by the torque ripple , which is the ratio of the torque deviation to the average of the output torque. The torque ripple is an important perfomance standard for the direct drive application, because the torque fluctuations prevent smooth motions, particularly at slow speeds, and degrade control accuracy.

To minimize the torque ripple, the rotor windings are divided into a number of coil pieces that are connected to the same number of commutator segments.   The current is then switched from one coil to another more smoothly. The number of poles at the stator is also increased in order to smoothen the magnetic field.

2.2.3 The Motor Constant

Another important performance standard is concerned with the power performance necessary for exerting torque. Let us consider a stall condition in which the motor does not rotate while exerting a torque. The power consumed inside the motor is evaluated by using Equs. (2.3) and (2.6), so that

$$P = Vi = \frac{R}{K_t}\tau \cdot \frac{\tau}{K_t} = \frac{R}{K_t^2}\tau^2 \qquad\qquad (2.7)$$

or

$$\tau = K_m\sqrt{P} \qquad\qquad (2.8)$$

where the constant K_m is given by

$$K_m = \frac{K_t}{\sqrt{R}} \quad (Nm/\sqrt{W}) \qquad\qquad (2.9)$$

As shown in Equ. (2.7) the output torque is proportional to the square root of the power consumed. The proportionality constant, K_m, accounts for the efficiency of the torque motor in terms of the transduction of electric power to mechanical torque. The higher the constant is, the larger the exerted torque becomes, and the lower the power consumed. The constant is referred to as the motor constant -- one of the major specifications of a direct-drive motor.

Let us also consider a steady-state condition when the motor is rotating at a constant angular velocity ω. From Equs. (2.2), (2.3), (2.5) and (2.8), we obtain

$$\tau = \frac{K_t}{R}V - K_m^2\omega \qquad\qquad (2.10)$$

Figure 2-1 plots the above torque-speed characteristics for a given armature voltage. The squared motor constant, K_m^2, designates the negative slope of the characteristics curve. The negative slope accounts for a natural damping effect, which results from the back emf . The negative slope contributes a stabilizing effect to the system.

When rotating at a constant speed ω, the motor produces the mechanical power P_m given by

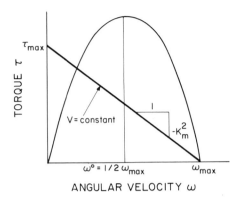

Figure 2-1: Torque-speed characteristics and power vs.
speed for direct current motor.

$$P_m = \tau\omega = \left(\frac{K_t}{R}V - K_m^2\omega\right)\omega \qquad (2.11)$$

which takes a maximum for a given armature voltage V. As shown in Figure 2-1, the power is maximum at half the maximum speed and at half the maximum torque. This provides the optimum operating condition, in which the motor can exert its maximum power. As the operating condition of the torque and speed departs from the optimum point, the efficiency declines along the parabolic curve. Note that for a larger motor constant, the optimal point moves towards the lower speed, since the maximun speed decreases for a fixed armature voltage V. Direct-drive motors are generally operated at low speeds, and in consequence a larger motor constant is desired in order to improve the efficiency.

2.2.4 The Electric Time Constant

Next we consider dynamic performances, particularly the response of the output torque to a sudden change of the armature voltage V. From Equs. (2.3) and (2.6), the Laplace transform of the motor torque is

$$\tau(s) = \frac{\dfrac{K_t}{R}V(s) - K_m^2\Omega(s)}{1 + T_e s} \qquad (2.12)$$

where T_e is referred to as the electric time constant and is given by

$$T_e = \frac{L}{R} \qquad (2.13)$$

The electric time constant is neglibly small for most D.C. motors, but it isn't for direct-drive torque motors. To generate a large torque, the direct-drive motor needs a large rotor diameter and larger windings, which possess, unavoidably, a large inductance.

It is of special importance for the motor drive system to cope with the significant effect of the large electric time constant so that the control system can be stable.

2.3 Brushless D.C. Torque Motors

2.3.1 Principle

The D.C. torque motors , investigated in the previous section, have many advantages including high efficiency , low torque ripple , linear torque-speed characteristics, and simplicity of construction.

A major drawback of the D.C. torque motor is that large currents are delivered to the rotor through mechanical commutation. At the brush and commutator , large sparks are created because of the large inductance at the rotor windings as well as the large currents and high voltages supplied to the rotor windings. The large sparks are harmful, causing the brush to wear quickly. Such a large spark also produces unwanted noise, which is harmful to other electrical devices. The brush mechanism increases mechanical friction, which degrades control performances.

The MIT Direct-Drive Arms and other commercialized robots employed brushless D.C. torque motors . In the brushless motor, the mechanical commutation is replaced by electric switching circuits, so that the mechanical brush can be eliminated. No sparks are then produced, while the features of the conventional D.C. motors are mostly preserved. The efficiency is as high as the conventional D.C. motors, and the torque-speed characteristics are linear in a wide range of speeds.

Unlike the conventional D.C. motors, the rotor of the brushless motor consists of permanent magnets, while the stator consists of windings (as shown in Figure 2-2). Thus the rotor and the stator are interchanged. The commutation of currents is accomplished by measuring the rotor position using a position sensor. In accordance with the rotor position, currents are delivered to appropriate windings through electronic switching circuits.

As mentioned in the previous section, the conventional D.C. torquer has many coil pieces and the same number of commutator segments, so that the torque ripple can be supressed. If, in the brushless motor, an individual electronic switch is used for each of the coil pieces, the drive amplifier becomes complex and costly, having many expensive switching transistors. Therefore, the number of independent switches is minimized in the brushless motor -- a common construction of the windings is that of a three-phase motor, as shown in Figure 2-3. This winding construction requires only independent switches, each of them is to draw a current into the winding at appropriate locations of the

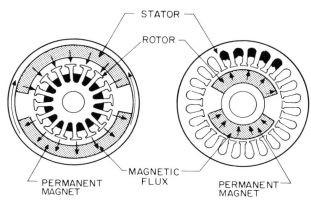

(a) Conventional DC motor   (b) Brushless DC motor

Figure 2-2: Constructions of brushless D.C. motor
and conventional D.C. motor.

Direct-Drive Robots

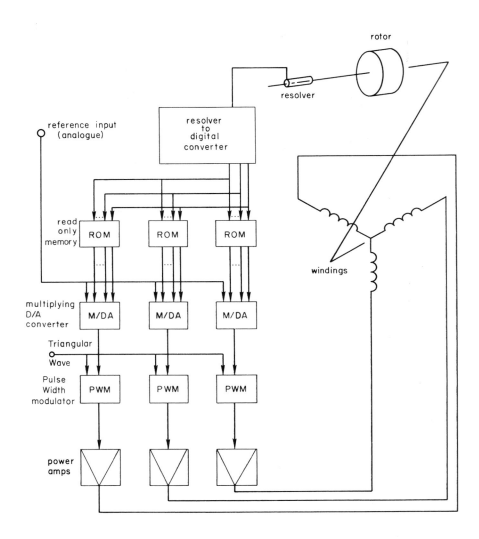

Figure 2-3: Brushless D.C. motor control system.

rotor.

A problem with this motor design is that the torque may change discontinuously when the switches are turned on and off as the rotor position changes. To reduce such a torque ripple , the current flowing into each winding is varied continuously with the rotor motion.

Let I_A, I_B and I_C be individual currents flowing into the three phase windings. Let them vary so that:

$$I_A = I_0 \sin \Theta_{rot}$$

$$I_B = I_0 \sin \left(\Theta_{rot} + \frac{2}{3}\pi \right)$$

$$I_C = I_0 \sin \left(\Theta_{rot} + \frac{4}{3}\pi \right)$$

(2.14)

where Θ_{rot} represents the rotor position and I_0 is some desired motor current. The torque exerted by the motor is given by

$$\tau = k_0 \left[I_A \sin \Theta_{rot} + I_B \sin \left(\Theta_{rot} + \frac{2}{3}\pi \right) + I_C \sin \left(\Theta_{rot} + \frac{4}{3}\pi \right) \right]$$

(2.15)

The first term accounts for the contribution by the windings of phase A, and the second and the third terms by the phase B and C windings, respectively. The motor torque τ is then obtained by substituting Equ. (2.14) into Equ. (2.15), and is given by

$$\tau = \frac{3}{2} k_0 I_0$$

(2.16)

The above equation indicates a linear relationship between the output torque and the motor current. In practice, this relationship may not be linear. The torque current characteristics of a brushless motor are virtually the same as those of a DC motor. Specifically, the torque constant and motor constant can be defined in the same way as was discussed previously. The choice of the phase currents given by Equ. (2.14) is implemented using ROM's and multiplying D/A converters as shown in Figure 2-3. A description of this implementation is given in Chapter 8. Another feature of brushless DC motors is their effectiveness in heat dissipation. This is because the windings are mounted on the stator which is in contact with the motor housing through a large area.

2.3.2 Commercialized Motors

Photo 2-2 shows commercialized brushless D.C. torque motors developed by International Scientific Inc.; the same type of motors as the ones used for MIT Direct-Drive Arms. The permanent magnets cemented on the rotor rings are high performance Samarium-Cobalt magnets , whose maximun magnetic energy product is over 26 MGOe. This rare earth magnet stores 3 to 10 times larger magnetic energy than conventional magnets, such as ferrite and AlNiCo magnets.

Table 2-1 shows specifications of one of the largest brushless motors taken from the manufacturer's catalogue. The maximum torque of the motor is 660 *Nm*, which is an order of magnitude larger than for a conventional D.C. torquer of the same size. The elimination of the brush allows us to supply large currents to the windings, and the strong magnets produce a high flux density. Since the magnetic flux is attenuated somehow by the currents flowing in the stator windings, the maximum torque listed in the catalogue data should be decreased. Nevertheless, this type of motor exerts the largest torque among the three types of the direct-drive motors.

A variety of techniques have been devised for reducing torque ripple. In the same way as conventional D.C. torquers, the number of poles is increased in the rotor design. As shown in Table 2-1, the 356 mm (14 inch) brushless motor has 30 poles. Increasing the number of slots on the stator is also effective to smoothen the torque.

Another technique to reduce the torque ripple is to arrange the rotor and stator pole positions to be ''allo-periodic''. If all the poles at the rotor and the stator are aligned at the same rotor position (iso-periodic), the resultant torque ripple becomes prominent. To avoid this, the number of the rotor poles is made slightly different from that of the rotor. The brushless D.C. torque motors in Photo 2-2 employ this technique to reduce the torque ripple.

The three-phase windings have been chosen for most brushless motors, mainly because the cost for switching transistors is minimized. If, however, the number of phases is increased to five or seven, it is effective to reduce the

Photo 2-2: Brushless D.C. torque motors.
(Courtesy of Shin Meiwa Industry Co., Japan)

Item	Size
Stator diameter (D)	$356\,mm$
Stator length (l)	$66\,mm$
Maximum torque	$660\,Nm$
Motor constant	$7.5\,Nm/\sqrt{W}$
Electrical time constant (T_e)	$10\,ms$
Weight	$12\,Kg.$

Table 2-1:   Specifications of brushless D.C. motor

torque ripple. Yet this method has not been adopted.

2.4 Variable Reluctance Motors

2.4.1 Principle

Another type of high-torque, low-speed motors used for direct-drive robots are variable reluctance motors . The principle of the variable reluctance motor is similar to that of a stepping motor, as shown in Figure 2-4. The stator has windings which create the flux passing through the stator and the tooth of the rotor, which is made of a magnetic substance. The rotor is at a stable position when the teeth of the rotor and the stator are aligned, as shown in Figure 2-4(a). A restoring force is generated as the rotor position varies from the stable position; Figure 2-4(b). Figure 2-4(c) shows the relationship between the restoring force f and the displacement Θ.

Figure 2-5 shows an extended structure of the motor in Figure 2-4. The stator has three phase windings; ϕ_1, ϕ_2, and ϕ_3. Note that the pitch of the rotor teeth is different from that of the stator. When the coil of phase ϕ_1 is excited, the rotor is stable at the position shown in the Figure. As the coil of phase ϕ_2 is excited consecutively, the rotor moves towards the tooth positions shown by the broken lines. As for the phase ϕ_3 coil, the stable position of the rotor is moved continuously by exciting the three-phase windings sequentially. Figure 2-5(b) shows the torque excited by the three-phase stepping motor.

The motor of this type is referred to as a Variable Reluctance Stepping Motor. The motor has some desirable advantages over the other types of

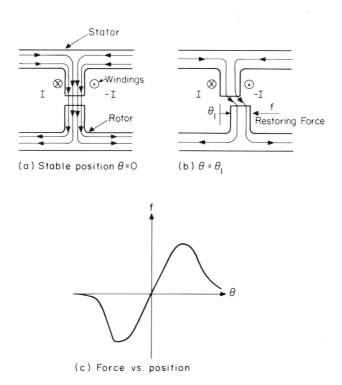

Figure 2-4: Force exerted in variable reluctance motor.

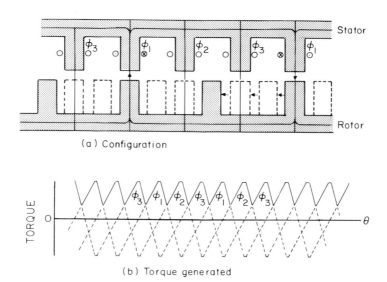

(a) Configuration

(b) Torque generated

Figure 2-5: Principle of variable reluctance stepping motor.

motors. However, to apply the variable reluctance motors to direct-drive robots, the following problems must be overcome:

1. When an excessively high speed or high acceleration motion is commanded, the rotor motion can be out of phase with the excitation timing of the stator windings.

2. The fluctuation in torque is prominent.

3. Since the rotor position is stable only when the teeth of the rotor are aligned with the ones of the stator, the possible positioning points are limited; namely coarse positioning resolution.

4. The ratio of motor weight to output torque tends to be higher than other motors.

To solve the first three problems, a closed loop current control using a rotor position sensor has been employed. With the rotor position information, the excitation of each phase windings can be accomplished correctively, thus no out-of-phase motion occurs.

As in the case of the brushless D.C. torquer, sinusoidal excitation currents can be produced with the rotor position information. In principle the rotor can be positioned at an arbitrary position by this closed-loop current control along with the sinusoidal phase excitation. The torque ripple can be also reduced on the same principle as the brushless motor .

2.4.2 Commercialized Motors

Efficient methods to increase the output torque are to reduce the air gap distance and to increase the number of teeth. Photo 2-3 shows the variable reluctance direct-drive motor developed by Motornetics Co., which employed the construction of increased teeth number. The number of teeth is 150 for the rotor and 144 for the stator. The motor consists of a rotor and the two stators; the inner and the outer stators as shown in Figure 2-6. This special construction was devised so that the drive torque is generated at both sides of the rotor. In fact, the number of teeth has increased to twice that of the single stator.

Each stator has 18 poles, and 6 segments of coils are connected in series. As shown in the figure, the windings are arranged so that the flux passes through the rotor in the radial direction. This construction allows the motor to reduce the internal loss of magnetic energy, and as a result a large motor constant can be obtained.

Table 2-2 shows major specifications of the 356 mm (14 inch) variable reluctance D.D. motor developed by Motornetics. Note that the motor constant

Photo 2-3: MEGATORQUE motor. (Courtesy of Motornetics).

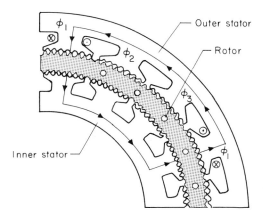

Figure 2-6: Construction of MEGATORQUE motor.

is larger than that of the brushless motor in Table 2-1. The maximum torque, however, is less than one half of that developed by the same size brushless motor. The continuous torque, on the other hand, is larger for the variable reluctance motor since the power dissipated in the variable reluctance motor is smaller than the brushless motor.

Variable reluctance motors , in general, have more prominant torque ripple and nonlinear torque-speed characteristics. Compensations for these nonlinearities are critical issues for their application to the direct-drive robots.

Item	Size
Stator diameter (D)	$356\,mm$
Stator length (L)	$102\,mm$
Maximum torque	$250\,Nm$
Motor constant	$23.2\,Nm/\sqrt{W}$
Coil inductance	$150\,mH$
Coil resistance	$3.5\,\Omega$
Weight	$60\,Kg.$

Table 2-2:   Specification of MEGATORQUE motor.

2.5 Issues in Arm Design and Control

Fundamental characteristics of the components involved in a direct-drive system were analyzed in the previous sections. We now discuss critical issues in the design and control of the direct-drive systems. The direct -drive system has several problems that make the application to the robot drive difficult. To exploit the potential features described in Chapter 1, we need to overcome the problems critical to the successful application to the robot drive.

2.5.1 Small Payload vs. Heavy Arm Weight

As noted in Section 1.3, the first implementation of the direct-drive robot at Carnegie-Mellon University resulted in a heavy arm weight and a small payload capacity. Having individual motors at the joints of the serial arm linkage, the robot becomes considerably heavy. The weight of the motor that drives a wrist joint is a load for the motor at the elbow. Both motors are then loads for the motor at a shoulder joint. Thus, the torque required for succeeding joints increases exponentially down the serial arm linkage. Accordingly, the size and weight of the succeeding motors increase exponentially [Asada and Kanade, 1983].

Alternative arm designs to suppress the exponential weight increase are necessary from a practical standpoint. As adopted in the M.I.T. Direct-Drive Arms and the commercialized arms, it is particularly efficient to use appropriate mechanisms for driving arm joints remotely with the motors located near the arm base. A critical issue of the remote drive is then to overcome some deficits resulting from the use of extra mechanical transmissions.

2.5.2 Damping Characteristics

The direct-drive system has almost zero damping , which may cause difficult problems in the control system design. Mechanical damping is produced only at the bearings supporting the joint shafts. The mechanical time constant , which is determined by the ratio of the inertial load to the mechanical damping coefficient, is then excessively large, compared with gear-drive systems. The back emf at the drive motor yields a damping effect, because the voltage applied to the motor is substantially reduced by the back emf, which is in proportion with the motor speed. This electromechanical damping effect is far less than the amount of damping necessary to stabilize the system response. In consequence, we need to increase the damping in the control system design.

Velocity feedback compensation is a straightforward approach to the improvement of the damping characteristics. As shown in Equ. (2.11), the damping coefficient increases by an amount $k_v K_t/R$ with velocity feedback. The control system design for the C.M.U. and M.I.T. arms revealed that one must choose the velocity feedback gain k_v to be a considerably large value in order to attain satisfactory responses. Accurate velocity measurement with a low noise level is then necessary to enable such a high gain velocity feedback. However, the velocity measurement for the direct-drive robot is more difficult than gear-drive robots. This is because the speed of a direct-drive motor is as slow as the link motion, while the motor speed of the gear-drive robot is much faster than the arm link, hence easier to measure the velocity. As a result, we need a high performance sensor appropriate for the measurment of the low speed motion.

2.5.3 Servo Stiffness

A robot manipulator arm is subject to a variety of disturbance forces particularly at its endpoint or the end effector. In positioning control, the robot is required to maintain the commanded position in the face of the external load. The ability to reflect a static disturbance force is often evaluated in terms of the endpoint stiffness [Asada and Slotine 86], that is, the ratio of the static force applied at the arm's endpoint to the resultant static deflection at the same point. The higher the endpoint stiffness becomes, the better the robot can bear the load and maintain the positioning accuracy .

The endpoint stiffness is determined by the feedback loop gains of individual joint servomechanisms as well as the mechanical construction of the manipulator arm. For a gear-drive robot, the loop gain increases with the gear ratio, n_i, since the motor output torque is amplified by the gear reducer. The direct-drive robot, however, cannot take this advantage of the reducer; there is no means to amplify the actuator torque mechanically. In consequence, the direct -drive robot needs a higher controller gain to compensate for the deficit of the loop gain . Again, the critical issue is the stability of the drive system.

2.5.4 Motor Inductance Effects

Direct-drive motors for robotics applications are generally designed for maximum output torques. To eliminate a mechanical gearing, the large output torque is the most important specification in the motor design. However, this requirement leads to the use of large windings, which have significantly large inductances. As described in Equ. (2.13), the electric time constant, which is determined by the motor inductance, becomes particularly slow for a large direct-drive motor having as large inductance. The slow time constant may cause a stability problem.

For traditional servomotors used in the gear-drive robot, the motor inductance is negligibly small. Hence, the effect of the electric time constant is negligible, and the drive system can be treated as a second- order system, unless other higher-order delays are prominant. Thus, the drive system is inherently stable with an appropriate damping. By contrast, often the direct-drive system can be unstable due to the motor inductance effect. To cope with this problem, the amplifier of the direct-drive motor must have fast dynamic responses to drive motor currents quickly in spite of the large inductance. Also, the control system must be designed carefully so that the stability margin can be sufficiently large.

2.5.5 Arm Dynamics Effects

In general, the dynamic behavior of a manipulator arm is highly coupled and nonlinear [Asada and Slotine 86]. The arm inertia varies depending upon the arm configuration. The acceleration of one joint affects other joints due to the reaction of the accelerated joint. This reaction causes coupling among the multiple joints. The configuration dependency of the manipulator inertia matrix induces the Coriolis and centrifugal torques, which are nonlinear torques comprising products of joint velocities.

When the manipulator arm is actuated with gearing, the complex arm dynamics are attenuated by the gearing when reflected to the motor shafts. Let n_i be the gear reduction ratio of the ith drive mechanism, then the dynamic equation of the ith motor is given by

$$\left(J_{rot} + \frac{J_{arm}}{n_i^2} \right) \ddot{\alpha}_i + \frac{\tau_{coup}}{n_i} + \frac{\tau_{non}}{n_i} = \tau_{mot} \tag{2.17}$$

where τ_{mot} is the torque exerted by the motor, and τ_{coup} and τ_{non} are, respectively, the coupling and nonlinear torques, which are attenuated by the factor of n_i when reflected to the motor axis. The first term of the above equation accounts for the inertial torque associated with the acceleration of the actuator, $\ddot{\alpha}_i$. The moment of inertia is comprised of the inertia of the motor rotor J_{rot}, including the reducer inertia, and the inertia of the arm links J_{arm}, which is reduced by the square of the gear ratio n_i when reflected to the motor axis. It should be noted that the arm link inertia varies depending upon the arm configuration, whilst the rotor inertia is invariant. The effect of the varying inertia is negligibly small upon the motor dynamics, if the gear ratio n_i is sufficiently large. The system parameters can be modeled as constants for the manipulator arm.

The direct-drive arm, where the gear ratio is equal to 1, is significantly different from the gear-drive robot. The complex arm dynamics are directly reflected to the motor axes due to the direct coupling. The varying inertia effect as well as the effects of the coupling and nonlinear torques are more prominant for the direct-drive robot. This creates difficult problems in the design of the control system. One should accomplish stable responses and sufficiently high gains despite the highly nonlinear and coupled dynamics. Considering the other substantial difficulties discussed above, we need to develop an efficient method to decouple the arm dynamics complexity from the control problems of the individual actuators.

2.5.6 Overheat

When a manipulator arm is at rest, the weight of the arm links is a load to each joint. For the gear-drive robot, the gravity load is attenuated by the gear reducer by a factor of n_i as shown in Equ. (2.17). Moreover, the gearing mechanism introduces friction , which acts in the opposite direction of the joint motion. As a result, the gravity load can be borne in part by the friction at the gearing.

As the gearing is eliminated in the direct-drive robot, the gravity load must be borne by the motor entirely and directly. This requires the motor to exert a larger amount of continuous torque for a long time. Thus, another critical issue in the direct-drive robot is the overheating problem under the continuous load condition. Efficient arm design techniques and motor technologies must be developed in order to avoid the overheat of the motors.

Part II

ARM DESIGN THEORY

Chapter 3

Statics

3.1 Introduction

In this chapter, we focus on the static characteristics of manipulators. Specifically, we consider a manipulator, such as the one shown in Figure 3-1, where n actuators are used to drive the arm linkage. Each actuator produces a torque τ which is combined with other motor torques through the link mechanism to generate a force $\mathbf{F}$ at the endpoint to do work. The ability to produce a useful force $\mathbf{F}$ without overheating the actuators is of great importance in the design of direct-drive arms, since all gearing has been eliminated. In order to study these characteristics, the system consisting of the drive mechanisms and the arm linkage must be considered. In what follows, an analytical tool is presented for evaluating mechanisms in terms of the power dissipation in the actuators and the end point force generated. This is particularly important in evaluating the overheating of direct-drive actuators under static load conditions. Two different mechanisms are selected and compared for use in the direct-drive approach.

3.2 Power Efficiency Analysis

One of the problems with the direct drive approach is that the robot cannot exert a large force. Because of the direct coupling of motors to their loads, the output torques of motors are not amplified through reducers. In consequence, the motors have to bear the loads directly, and can overheat, particularly in a stall condition. In this section, we analyze the power dissipated in the motors when a force is exerted at the tip of the arm.

An important parameter of torque motors is the motor constant as described in Chapter 2 and given by Equ. (2.8). It is the ratio of the output torque τ to the square root of power dissipated P in the motor [Electro-Craft 80],

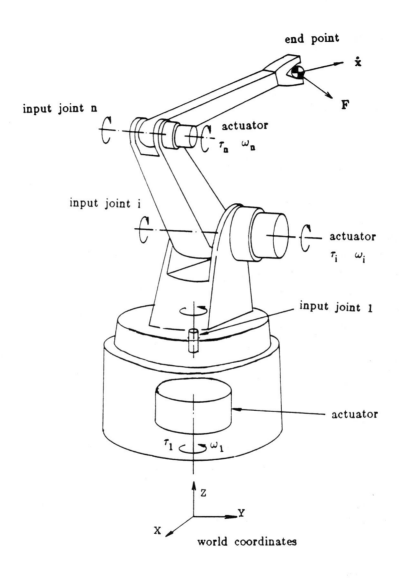

Figure 3-1:   A robot manipulator

$$k_m = \frac{\tau}{\sqrt{P}} \qquad (3.1)$$

The motor constant represents the efficiency with which the motor converts its input power into output torque. If k_m is small, the motor dissipates a large amount of power to exert a small torque.

When a robot bears a load at the arm tip, the load is distributed among all the motors of the arm mechanism. Hence, in addition to the individual motor characteristics, the efficiency of exerting a force at the arm tip also depends on the kinematic structure of the arm mechanism. Let **F** be the end point force exerted at the arm tip represented in cartesian coordinates, and let τ be the vector of equivalent torques of n motors, then one obtains [Paul 81]

$$\tau = \mathbf{J}^T \mathbf{F} \qquad (3.2)$$

where **J** is the jacobian matrix associated with the coordinate transformation from the joint coordinates to the cartesian coordinates of the arm tip. A brief note on the jacobian matrix is given in Appendix D.

Let P be the total power dissipated in the n motors driving the arm mechanism and k_{mi} be the motor constant of the i-th motor, then the total power dissipation, P, is given by the following quadratic form of end point force **F**,

$$P = \sum_{i=1}^{n} \frac{\tau_i^2}{k_{mi}^2} = \mathbf{F}^T \mathbf{L} \mathbf{F} \qquad (3.3)$$

where **L** is given by

$$\mathbf{L} = \mathbf{J} \, diag\left(\frac{1}{k_{m1}^2}, \ldots, \frac{1}{k_{mn}^2}\right) \mathbf{J}^T \qquad (3.4)$$

Thus the power dissipation depends not only on the kinematic structure, but also on the motor constants of the actuators.

The relation given by Equ. (3.4) is a congruence transformation, and thus the matrix **L** is positive definite as long as the Jacobian matrix is nonsingular [Strang 80]. Therefore, the matrix **L** has all real, positive eigenvalues, $\lambda_i > 0$, and the quadratic form can be rewritten as

$$P = \sum_{i=1}^{n} \lambda_i F_i^2 \qquad (3.5)$$

where F_i is the component of end point force along the eigenvector corresponding to λ_i. The geometrical interpretation of Equ.(3.3) is obtained by fixing the total power P. In this case , the quadratic form of Equ. (3.3) is that of an ellipsoid as shown in Figure 3-2. The principal axes of such an ellipsoid are aligned with the eigenvectors of the matrix **L**, and the length of each principal axis is equal to the reciprocal of the square root of the corresponding eigenvalue. The static characteristics will depend on arm configuration since the smallest (largest) eigenvalue of **L** corresponds to the major (minor) axis of the ellipsoid, the maximum (minimum) force **F** acts along the major (minor) axis. Therefore, along the major axis, the input power is most effectively converted into a force at the arm tip. Since the matrix **L** involves the Jacobian matrix **J**, which varies with the arm configuration. By examining the ellipsoid for different arm configurations, one can analyze the global characteristics of the power to force conversion.

To evaluate the overall   efficiency of the arm-motor combination, we define the   mean power dissipation ratio, λ_m, which is the mean of the eigenvalues and is given by

$$\lambda_m = \frac{1}{n} trace(\mathbf{L}) \tag{3.6}$$

The units of λ_m are Watts per square Newton.

3.3 Kinematic Structure Design

3.3.1 Parallel Drive Mechanism

In order to select a suitable kinematic structure for a direct-drive arm, we consider the following basic structures : a simple serial drive mechanism, as shown in Figure 3-3; and a general five-bar-link parallel drive mechanism, shown in Figure 3-4.

In this section we define these two drive system methods: serial drive and parallel drive. Figure 3-3 shows the typical construction of a serial drive arm mechanism, in which the lower link is driven by a motor fixed on the base and the upper link is driven by a motor attached at the tip of the lower link. In this serial configuration, the weight of the second motor is a load to the first motor. Moreover, the   reaction torque of the second motor acts on the first motor.

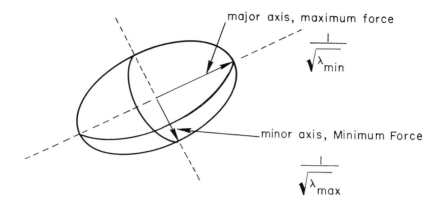

Figure 3-2: Force ellipsoid.

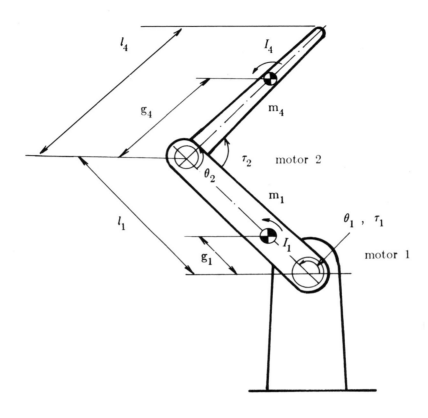

Figure 3-3:   2 d.o.f serial drive arm mechanism.

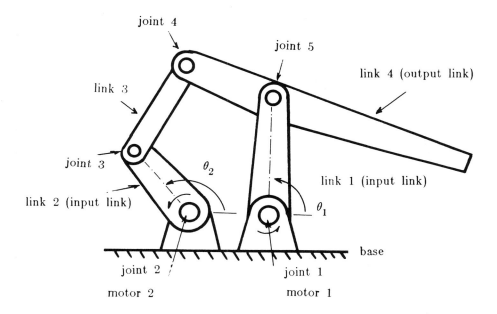

Figure 3-4: A general five-bar-link mechanism

When the second motor accelerates in the clockwise direction, a counter-clockwise torque acts on the first motor, and vise versa. Thus the two motors have significant interactions. In this book, an arm mechanism in which each motor is located at a joint of a serial linkage between the adjacent links is referred to as a serial drive mechanism.

Figure 3-4 shows an alternative arm construction consisting of a five-bar-link mechanism. Two motors fixed on the base drive the two input links and cause a two-dimensional motion at the tip of the arm. The weight of one motor is not a load on the other. Also the reaction torque of one motor does not act directly upon the other, because the motors are fixed on the base . An arm mechanism in which motors are mounted on a fixture and the weight and reaction torque of one motor do not affect the other motors directly is referred to as a parallel drive mechanism. The parallel drive mechanism has been used in several models of commercial robots that have conventional electromechanical drives with gearings. The five-bar-link mechanism, however, is particularly useful for overcoming the inherent difficulties of the direct drive method.

3.3.2 Motor Constant Identification

In order to use the analysis of the previous section, the motor constants, k_{mi}, have to be determined. The following experiment was conducted to identify the motor constant of a single active joint. First, a position control system is set up to permit the arbitrary positioning of the motor rotor using a computer terminal (see Chapter 8). This position control system is needed because the relationships derived in the previous section are valid at stall condition. Second, known torque loads were applied to the motor shaft while under position control and the power dissipated by the system was calculated using the phase currents in the motor. These phase currents are in error due to sensor limitations and must be corrected. A procedure in correcting these current readings is given in Appendix A. Figure 3-5 provides a plot of the experimental data. The square root of power measured is plotted against the torque applied. A least square fit yields a motor constant of $2.366\,Nm/\sqrt{Watts}$.

3.3.3 Kinematic Structure Evaluation

The end point force for the serial drive mechanism in comparison with the parallel drive mechanism of Figure 3-6 is evaluated. These characteristics were analyzed in [Asada and Youcef-Toumi 83a, Asada and Youcef-Toumi 84].

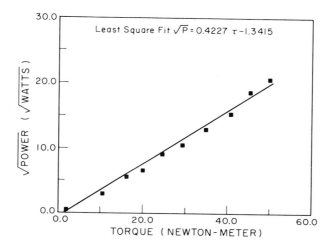

Figure 3-5: Motor constant.

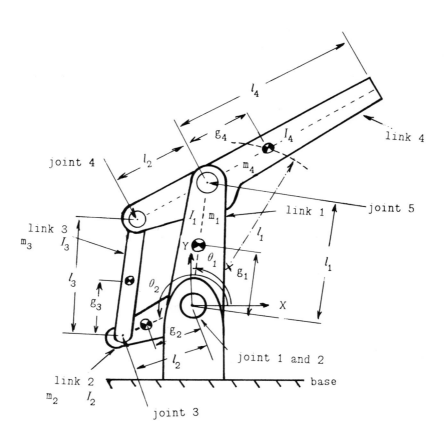

Figure 3-6: 2 d.o.f parallel drive five-bar-link mechanism

For the five-bar-link parallel drive mechanism, the matrix **L** is given by

$$\mathbf{L} = \begin{bmatrix} L_{11} & L_{12} \\ L_{21} & L_{22} \end{bmatrix}$$

where

$$L_{11} = \frac{l_1^2}{k_{m1}^2} \sin^2(\theta_1) + \frac{l_4^2}{k_{m2}^2} \sin^2(\theta_2)$$

$$L_{21} = L_{12} = -\frac{l_1^2}{k_{m1}^2} \sin(\theta_1)\cos(\theta_1) - \frac{l_4^2}{k_{m2}^2} \sin(\theta_2)\cos(\theta_2)$$

$$L_{22} = \frac{l_1^2}{k_{m1}^2} \cos^2(\theta_1) + \frac{l_4^2}{k_{m2}^2} \cos^2(\theta_2)$$

where the end point force **F** is represented in the coordinate system fixed at the base as shown in Figure 3-6.

Figure 3-7 illustrates the static force characteristics, Equ. (3.3), for the parallel configuration. The maximum (minimum) force produced is along the major (minor) axis of the ellipsoid at the tip of the arm. Note that in this parallel configuration the force appears to be nearly isotropic in a wide region . The maximum force based on a total power of $1KW$ is estimated to be $190N$. This value is obtained by taking the square root of the ratio of the power to the minimum eigenvalue.

The mean power dissipation ratio is then given by

$$\lambda_m = \frac{1}{2}\left(\frac{l_1^2}{k_{m1}^2} + \frac{l_4^2}{k_{m2}^2} \right) \tag{3.7}$$

Although the matrix **L** varies with the arm configuration, the mean power dissipation λ_m is uniform for all the configurations. For shorter link lengths, l_1 and l_4, and larger motor constants, k_{m1} and k_{m2}, the arm-motor combination has a better efficiency having a small mean power dissipation ratio.

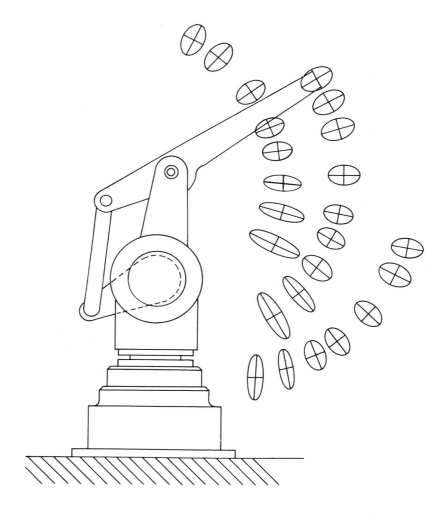

Figure 3-7:   Static force characteristics

Let us compare the parallel drive mechanism with the serial drive in terms of the value of λ_m. The mean power dissipation ratio of the serial drive mechanism of Figure 3-3 is

$$\lambda_m = \frac{1}{2}\left(\frac{l_1^2}{k_{m1}^2} + \frac{l_4^2}{k_{m2}^2} + \frac{l_4^2}{k_{m1}^2}\right) - \frac{l_1 l_4}{k_{m1}^2}\cos(\theta_2) \tag{3.8}$$

In this case, λ_m varies depending on the arm configuration. As θ_2 increases, λ_m becomes larger, namely, the efficiency becomes lower. Comparing Equ. (3.7) with Equ. (3.8), one finds the parallel drive has a smaller mean power dissipation ratio than the serial drive except when

$$|\theta_2| \leq \cos^{-1}\left(\frac{l_4}{2l_1}\right) \tag{3.9}$$

The mean power curves are shown in Figure 3-8.

In the case of the M.I.T. direct-drive robot Model II described in Figure 3-6, link lengths, l_1 and l_4, are $0.40\,m$ and $0.65\,m$ respectively, and the angle of joint 2 is limited by $35° \leq \theta_2 \leq 170°$ The parallel drive mechanism has a better efficiency for all arm configurations within the limit. This is another important advantage of the parallel drive mechanism over the serial drive. Having the same motor constants and the same link lengths throughout the same workspace, the parallel drive mechanism can more efficiently exert force at the arm tip.

Let us now consider the same mechanism, where the links are crossed, thus yielding completely different kinematic characteristics. This cross drive configuration (mode II) depicted in Figure 3-9, shows a complete change in the static force characteristics. The force ellipsoid reduces to a line in the reachable region. The ratio between the maximum and minimum force at any point in space is greater than 11. This value is on the order of 1.6 for the parallel drive of MODE I. The MODE II configuration of this mechanism can thus be used in applications where a large force is desired along the directions shown in Figure 3-9. This force, while having a preferred direction, is much larger than what the mechanism in MODE I configuration can produce. The maximum payload based on a total power of 1 KW is $33\,Kg$. The M.I.T. direct-drive robot models II and IV have the ability to change modes [Asada and Youcef-Toumi 83a].

Direct-Drive Robots

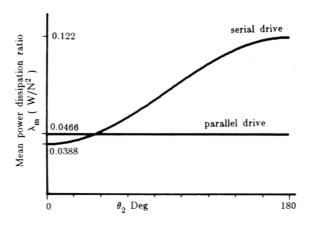

Figure 3-8: Mean power dissipation ratio for the parallel
and serial drives.

The direct-drive robot model IV is shown in these two different modes in Photos 3-1 and 3-2. Article 2 of Part IV uses these ideas to design linkages for direct- drive arms.

3.4 Conclusion

This chapter has presented an analytical tool for evaluating actuated mechanisms in terms of power dissipation and endpoint force static characteristics. These tools are shown to be useful in selecting an appropriate arm linkage design for a specific application. As an example, by analyzing the internal power dissipation in the motors when the manipulator exerts a force at the endpoint, it was shown that the parallel drive mechanism has a lower power dissipation than a serial drive that uses the same motors and covers the same workspace.

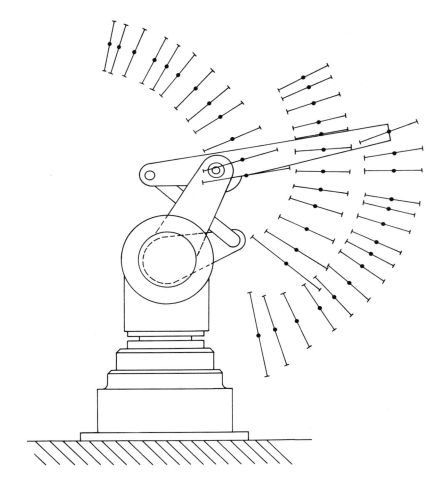

Figure 3-9: Static force characteristics for the
cross configuration (mode II)

Photo 3-1: M.I.T. direct-drive robot model IV in parallel drive mode

Photo 3-2: M.I.T. direct-drive robot model IV in cross drive mode.

Chapter 4

Arm Design for Simplified Dynamics

4.1 Introduction

A manipulator design theory for reduced dynamic complexity is presented. The kinematic structure and mass distribution of a manipulator arm are designed so that the inertia matrix in the equation of motion becomes diagonal and/or invariant for an arbitrary arm configuration. For the decoupled and invariant inertia matrix, the system can be treated as a linear, single-input, single-output system with constant parameters. Consequently, control of the manipulator arm is simplified, and more importantly, the reduced dynamic complexity permits improved control performance.

First, the problem of designing such an arm with a decoupled and/or configuration-invariant inertia matrix is defined. The inertia matrix is then analyzed in relation to the kinematic structure and mass properties of the arm links. Necessary conditions for a decoupled and/or configuration-invariant manipulator inertia matrix are then obtained. Using the necessary conditions, the kinematic structure and mass properties are found which reduce the inertia matrix to a constant diagonal form. Possible arm designs for decoupled and/or invariant inertia matrices are then determined for 2 and 3 degree-of-freedom manipulators.

Dynamic complexity, such as coupling and nonlinearities are major concerns in the control of manipulator arms. This problem is critical in manipulators designed for high speed applications which have particularly prominant dynamic complexity. Also, in direct-drive arms and other manipulator arms in which the gear ratios of reducers are low, the complicated dynamic has a more direct influence upon the drive mechanisms [Asada and Youcef-Toumi 84]. Again the control problem becomes more complex and difficult because of the prominant coupling and nonlinearities.

A number of control methods have been developed for manipulators. Applications of model-referenced adaptive control [Dubowsky and Des Forges 79], [Takegaki and Arimoto 81], nonlinear feedback control [Freund 82], and sliding mode control [Utkin 72, Utkin 77, Utkin 78, Young 78, Slotine and Sastry 83] have made significant contributions to the improvement of control performance. Dynamic computation algorithms [Luh, Walker, and Paul 80], [Hollerbach 80], parameter estimation methods [Atkeson, An, and Hollerbach 85], [Mayeda, Osuka, and Kangawa 84], and customized inverse dynamics methods [Kanade, Khosla and Tanaka 84] have allowed us to compensate for the complicated arm dynamics in real time and in an intelligent manner.

Another method for reducing dynamic complexity and improving dynamic performance involves the coordination of the controller and manipulator design efforts. In this method, the mechanical construction of a manipulator arm is modified so that the resultant dynamic behavior can be improved or becomes desirable for control [Asada 83]. In general, the dynamics of a manipulator arm depends upon the mass properties of individual arm links and the kinematic structure of the arm linkage. Redistribution of mass and modification of the arm structure can provide improvements in the manipulator dynamics.

The goal of this chapter is to explore arm linkage design for reducing dynamic complexity, and consequently reducing the control difficulties. First, some particular forms of the dynamic equations are presented in which the inertia matrix is reduced to a diagonal and/or configuration invariant form. Then, a design theory is developed for acheiving the desired manipulator dynamics. Finally, possible arm structures and mass distributions for reducing the dynamic complexity are determined.

4.2 Decoupling and Configuration-Invariance of Manipulator Dynamics

Desirable forms for the manipulator dynamics are formulated in this section. The significance of each of these forms is discussed with reference to how the control problem can be simplified, and how much the computational complication is relaxed.

Let θ_i and τ_i be the joint displacement and torque of the ith joint, respectively, then the equation of motion of the manipulator arm is given by [Asada and Youcef-Toumi 84] [Asada and Slotine 86]

$$\tau_i = H_{ii}\ddot{\theta}_i + \sum_{j \neq i} H_{ij}\ddot{\theta}_j + \sum_{j}\sum_{k}\left(\frac{\partial H_{ij}}{\partial\theta_k} - \frac{1}{2}\frac{\partial H_{jk}}{\partial\theta_i}\right)\dot{\theta}_j\dot{\theta}_k + \tau_{gi} \qquad (4.1)$$

where H_{ij} is the $i{-}j$ element of the manipulator inertia matrix, and τ_{gi} is the torque due to gravity. The first term on the right hand side represents the inertia torque generated by the acceleration of the ith joint, while the second term is the interactive inertia torque caused by the accelerations of the other joints. The interactive inertia torque is linearly proportional to acceleration. The third term represents the nonlinear velocity torques resulting from Coriolis and centrifugal effects. In general, the dependence of the inertia matrix on the arm configuration produces these nonlinear velocity torques.

The inertia matrix is determined by the kinematic structure of the manipulator arm and the mass properties of individual links. The design problem discussed in this paper requires modification of both the kinematic structure and the mass properties of the arm linkage so that the inertia matrix reduces to the particular desired forms.

Consider the inertia matrix that reduces to a diagonal matrix for an arbitrary arm configuration, then the second term in Equ. (4.1) vanishes and no interactive torques appear. The manipulator inertia matrix in this case is referred to as a decoupled inertia matrix. The significance of the decoupled inertia matrix is that the control system can be treated as a set of single-input, single-output subsystems associated with individual joint motions. The equation of motion under these conditions reduces to

$$\tau_i = H_{ii}\ddot{\theta}_i + \sum_{k}\left(\frac{\partial H_{ii}}{\partial\theta_k}\dot{\theta}_i\dot{\theta}_k - \frac{1}{2}\frac{\partial H_{kk}}{\partial\theta_i}\dot{\theta}_k^2\right) + \tau_{gi} \qquad (4.2)$$

where the second term represents the nonlinear velocity torques resulting from the spatial dependency of the diagonal elements of the inertia matrix. Note that the number of terms involved in Equ. (4.2) is much smaller than the number of original nonlinear velocity torques, because all the off-diagonal elements are zero for all $\theta_1, \ldots, \theta_n$. This reduces the computational complexity of the nonlinear torques.

Another significant form of the inertia matrix that reduces the dynamic complexity is a configuration-invariant form. The inertia matrix in this case does not vary for an arbitrary arm configuration. In other words, the matrix is independent of joint displacements, hence the third term in Equ. (4.1) vanishes and the equation of motion reduces to

$$\tau_i = H_{ii}\ddot{\theta}_i + \sum_{j \neq i} H_{ij}\ddot{\theta}_j + \tau_{gi} \tag{4.3}$$

Note that the coefficients H_{ii} and H_{ij} are constant for all arm configurations. Thus, the equation is linear except the last term, that is, the gravity torque. The inertia matrix in this form is referred to as an invariant inertia matrix. The significance of this form is that linear control schemes can be adopted, which are much simpler and easier to implement.

When the inertia matrix is both decoupled and configuration invariant, the equation of reduces to

$$\tau_i = H_{ii}\ddot{\theta}_i + \tau_{gi} . \tag{4.4}$$

The system is completely decoupled and linearized, except the gravity term. Thus, we can treat the system as single-input, single-output systems with constant parameters.

The decoupled and configuration-invariant inertia matrix was first accomplished in reference [Asada and Youcef-Toumi 84], in which a special five-bar-link parallel drive mechanism was devised for a direct-drive arm. The feasibility and practical usefulness were also demonstrated through the development of the special arm. However, extension to general manipulator arms was not made in a comprehensive form. In this chapter, the conditions for the kinematic structure and mass properties of the manipulator arm which provide a decoupled and/or configuration-invariant inertia are determined.

4.3 Modelling

In this section, we derive the inertia tensor matrices of the manipulator arms discussed in this paper. As shown in Figure 4-1, the manipulator arm is assumed to be an open kinematic chain consisting of only revolute joints. The

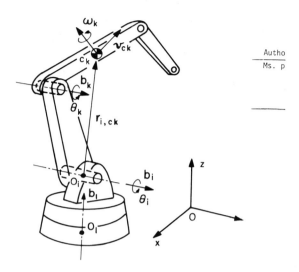

Figure 4-1: An open kinematic manipulator with revolute joints.

joints are numbered 1 through n from the proximal joint to the distal joint. The link between joints i and $i+1$ is called link i. The direction of the axis of joint i is represented by a unit vector $\mathbf{b}_i$ and the displacement of link i is denoted by θ_i which is the angle of rotation about the unit vector. The centroid of link k is shown by point c_k in the figure, the velocity vector of the centroid is denoted by $\mathbf{v}_k$ and the angular velocity vector by ω_k. Let m_k and I_k be the mass and the inertia tensor of link k with respect to the $O-xyz$ inertial reference frame, then the total kinetic energy stored in the arm links from 1 to n is given by

$$T = \sum_{k=1}^{n} \frac{1}{2} \left(m_k \mathbf{v}_{ck}^T \mathbf{v}_{ck} + \omega_k^T I_k \omega_k \right) \tag{4.5}$$

The motion of link k is generated by the preceding joint motions. The angular velocity ω_k, for example, is given by

$$\omega_k = \sum_{i=1}^{k} \mathbf{b}_i \dot{\theta}_i \tag{4.6}$$

To represent the linear velocity of the centroid c_k, we denote the position vector from an arbitrary point on the ith joint axis to the centroid c_k by vector $\mathbf{r}_{i,ck}$. Then,

$$\mathbf{v}_{ck} = \sum_{i=1}^{k} \mathbf{b}_i \dot{\theta}_i \times \mathbf{r}_{i,ck} \tag{4.7}$$

where $\times$ is the vector product. Substituting Equ. (4.6) and Equ. (4.7) into Equ. (4.5) yields

$$T = \frac{1}{2} \sum_{i=1}^{n} \sum_{j=1}^{n} H_{ij} \dot{\theta}_i \dot{\theta}_j \tag{4.8}$$

where H_{ij} is the $i-j$ element of the $n \times n$ manipulator inertia matrix involved in Equ.(4.1), and is given by

$$H_{ij} = \sum_{k=\max[i,j]}^{n} \left[m_k \left(\mathbf{b}_i^T \mathbf{b}_j \cdot \mathbf{r}_{i,ck}^T \mathbf{r}_{j,ck} - \mathbf{b}_j^T \mathbf{r}_{i,ck} \cdot \mathbf{b}_i^T \mathbf{r}_{j,ck} \right) + \mathbf{b}_i^T I_k \mathbf{b}_j \right] \tag{4.9}$$

Note that the inertia matrix is symmetric, hence $H_{ij} = H_{ji}$. We also derive a different expression for the inertia matrix, which we need in the following analysis. Let l be an arbitrary joint number. We consider the case in which all the joints between $i+l$ and n are immobilized. The last $(n-l+1)$ links are then treated as a single rigid body. As shown in Figure 4-2, let $M_{l,n}$ and $\mathbf{p}_l$ be, respectively, the total mass of the last $(n-l+1)$ links and the position vector from point O_l to the centroid $C_{l,n}$ that is the centroid of the last $(n-l+1)$ links. Then,

$$M_{l,n} = \sum_{k=l}^{n} m_k \tag{4.10}$$

$$\mathbf{p}_l = \sum_{k=l}^{n} \frac{m_k \mathbf{r}_{l,ck}}{M_{l,n}} \tag{4.11}$$

Let $\mathbf{q}_{l,k}$ be the position vector from the centroid $C_{l,n}$ to the centroid of link k, then the following is also derived from the definition of the centroid:

$$\mathbf{r}_{i,ck} = \mathbf{r}_{i,l} + \mathbf{p}_l + \mathbf{q}_{l,k} \tag{4.12}$$

$$\sum_{k=1}^{n} m_k \mathbf{q}_{l,k} = 0 \tag{4.13}$$

Substituting Equ. (4.12) and Equ. (4.13) into Equ. (4.9), we can split the i–j element of the inertia matrix into two parts: one involves terms associated with the last $(n-l+1)$ links and the other involves the terms for the other links.

$$H_{ij}^{(l)} = \sum_{k=\max[i,j]}^{l-1} \left[m_k \left(\mathbf{b}_i^T \mathbf{b}_j \cdot \mathbf{r}_{i,ck}^T \mathbf{r}_{j,ck} - \mathbf{b}_j^T \mathbf{r}_{i,ck} \cdot \mathbf{b}_i^T \mathbf{r}_{j,ck} \right) + \mathbf{b}_i^T \mathbf{I}_k \mathbf{b}_j \right] \tag{4.14}$$

$$+ \mathbf{b}_i^T \mathbf{N}_{l,n} \mathbf{b}_j + M_{l,n} \left[\mathbf{b}_i^T \mathbf{b}_j (\mathbf{r}_{i,l} + \mathbf{p}_l)^T (\mathbf{r}_{j,l} + \mathbf{p}_l) - \right.$$

$$\left. \mathbf{b}_i^T (\mathbf{r}_{i,l} + \mathbf{p}_l) \cdot \mathbf{b}_i^T (\mathbf{r}_{j,l} + \mathbf{p}_l) \right]$$

where $\mathbf{N}_{l,n}$ is the composite inertia tensor of the last $(n-l+1)$ links with respect

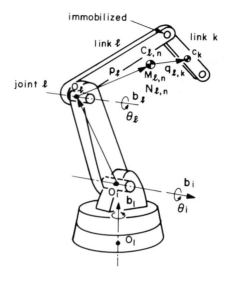

Figure 4-2: Total mass and centroid of last $(n-1+l)$ links.

to the centroid $C_{l,n}$, which is given by

$$\mathbf{N}_{l,n} = diag \left(\sum_{k=1}^{n} m_k \mathbf{q}_{l,k}^T \mathbf{q}_{l,k} \right) + \sum_{k=1}^{n} \left(\mathbf{I}_k - m_k \mathbf{q}_{l,k} \cdot \mathbf{q}_{l,k}^T \right) \qquad (4.15)$$

4.4 Analysis

In order to eliminate the coupling and nonlinear torques, the inertia matrix must be diagonalized and made invariant for all the arm configurations. In this section, we analyze the inertia matrix and derive the necessary conditions for the kinematic structure and mass properties to allow the reduction of the inertia matrix to a diagonal and/or invariant form.

4.4.1 Off-diagonal elements

We first discuss the necessary conditions for an off-diagonal element, $H_{ij} = H_{ji}$ ($i<j$), to be zero or invariant for all the arm configurations. We derive the necessary conditions by considering a special situation where only one joint, say joint l, rotates 360 degrees, while the other joints are immobilized. This simple analysis provides useful conditions that are used to determine the kinematic structure and mass properties for decoupled and/or configuration-invariant inertia.

As shown in Figure 4-3, we immobilize all the joints beyond joint l, and regard the last $(n-l+1)$ links as a single rigid body. Let $OC_{l,n}$ be the perpendicular from the centroid $C_{l,n}$ to the joint axis $\mathbf{b}_l$. We locate the inertial reference frame $O-xyz$ at the point O, which is chosen to coincide with point O_l on joint axis $\mathbf{b}_l$. Without loss of generality, we can orient the z axis in the same direction as that of joint axis $\mathbf{b}_l$. The displacement of joint l is then represented by the angle from the x axis to the vector $\mathbf{p}_l$, which is the position vector of the centroid $C_{l,n}$ relative to the origin O. Denoting the length of the vector $\mathbf{p}_l$ by $L_{l,n}$, we obtain,

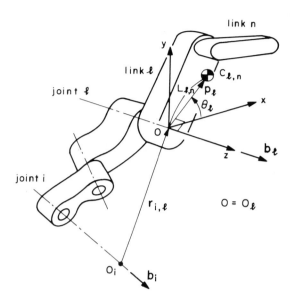

Figure 4-3: Immobilizing joints of a manipulator

$$\mathbf{p}_l = \begin{bmatrix} L_{l,n}\cos\theta_l \\ L_{l,n}\sin\theta_l \\ 0 \end{bmatrix} \tag{4.16}$$

The inertia tensor $\overline{\mathbf{N}}_{l,n}$, that is the composite inertia tensor of the last $(n-l+1)$ links, varies in its matrix expression, as the joint angle θ_l changes. Let $\mathbf{N}_{l,n}$ be the composite inertia tensor when $\theta_l = 0$, the matrix $\mathbf{N}_{l,n}$ is given by

$$\mathbf{N}_{l,n} = \mathbf{A}\overline{\mathbf{N}}_{l,n}\mathbf{A}^T \tag{4.17}$$

where $\mathbf{A}$ is the 3×3 rotation matrix given by

$$\mathbf{A} = \begin{bmatrix} \cos\theta_l & -\sin\theta_l & 0 \\ \sin\theta_l & \cos\theta_l & 0 \\ 0 & 0 & 1 \end{bmatrix} \tag{4.18}$$

For simplicity, the subscripts l and n are neglected unless the notation gets confusing. The elements of the inertia tensor $\mathbf{N}$ are then written as

$$\mathbf{N} = \begin{bmatrix} \overline{N}_{xx} & \overline{N}_{xy} & \overline{N}_{xz} \\ \overline{N}_{xy} & \overline{N}_{yy} & \overline{N}_{yz} \\ \overline{N}_{xz} & \overline{N}_{yz} & \overline{N}_{zz} \end{bmatrix} \tag{4.19}$$

When the joint axis $\mathbf{b}_i$ is not orthogonal to the joint axis $\mathbf{b}_l$, the ith joint axis intersects with the x–y plane. If we define the intersection as the representative point on the ith joint axis, the position vector $\mathbf{r}_{i,l}$, that is the vector from points O_i to O_l, can be written as

$$\mathbf{r}_{i,l} = \begin{bmatrix} r_x \\ r_y \\ r_z \end{bmatrix} \tag{4.20}$$

The direction cosines of the ith joint axis are denoted by

$$\mathbf{b}_i = \begin{bmatrix} b_x \\ b_y \\ b_z \end{bmatrix} \tag{4.21}$$

Substituting Equ. (4.16)-Equ. (4.21)) into Equ. (4.14) yields the expression relating the off-diagonal element of the inertia matrix H_{il} to the joint angle θ_l,

$$H_{il}^{(l)} = (ML^2 + \overline{N}_{zz})b_z + (b_x\overline{N}_{xz} + b_y\overline{N}_{yz} + MLb_z r_x)\cos\theta + \tag{4.22}$$
$$(b_y\overline{N}_{xz} - b_x\overline{N}_{yz} + MLb_z r_y)\sin\theta$$

Note that the subscripts l and n for M and L are neglected. The following proposition is derived directly from Equ. (4.22)

Proposition (4.1): The necessary condition for the off-diagonal element of the inertia matrix, H_{il}, to be invariant for an arbitrary θ_l are given by

$$b_x\overline{N}_{xz} + b_y\overline{N}_{yz} + MLb_z r_x = 0 \tag{4.23}$$

and

$$b_y\overline{N}_{xz} - b_x\overline{N}_{yz} + MLb_z r_y = 0 \tag{4.24}$$

In addition to the above conditions, the first term in Equ. (4.22) must be zero when the decoupled inertia matrix is required. However, the off-diagonal element H_{il} does not vanish unless $b_z = 0$, since the diagonal element of the inertia tensor $\overline{N}_{zz}$ is always positive. When $b_z = 0$, the two joint axes are perpendicular to each other, and we conclude the following.

Proposition (4.2): For manipulator arms with an open kinematic chain structure, the inertia matrix cannot be decoupled unless the joint axes are orthogonal to each other.

When $\mathbf{b}_i^T \mathbf{b}_l = 0$, namely the two joint axes are orthogonal, we can direct the x axis of the inertial reference frame along the joint axis $\mathbf{b}_i$, without loss of generality. As shown in Figure 4-4,

$$
\mathbf{b}_i = \begin{bmatrix} 1 \\ 0 \\ 0 \end{bmatrix}, \quad \mathbf{b}_l = \begin{bmatrix} 0 \\ 0 \\ 1 \end{bmatrix} \tag{4.25}
$$

Let O_i be the representative point of joint i that is located at the intersection of the joint axis and the y–z plane as shown in the figure, then the position vector $\mathbf{r}_{i,l}$ can be expressed as

$$
\mathbf{r}_{i,l} = \begin{bmatrix} 0 \\ r_y \\ r_z \end{bmatrix} \tag{4.26}
$$

Using the above equations in Equ. (4.14), we obtain

$$
H_{il}^{(l)} = (\overline{N}_{xz} - MLr_z)\cos\theta - \overline{N}_{yz}\sin\theta \tag{4.27}
$$

The following is derived directly from the above results

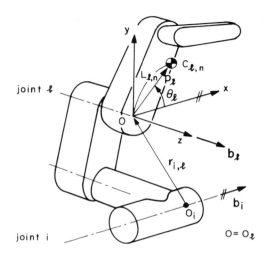

Figure 4-4: Manipulator with two orthogonal joint axes

Proposition (4.3): When joint axes $\mathbf{b}_i$ and $\mathbf{b}_l$ are orthogonal, the necessary conditions for the off-diagonal element H_{il} to vanish for an arbitrary joint angle θ_l are given by

$$MLr_z = \overline{N}_{xz} \tag{4.28}$$

and

$$\overline{N}_{yz} = 0 \tag{4.29}$$

4.4.2 Diagonal elements

Next, we consider diagonal elements of the inertia matrix, H_{ii}. Let l be an arbitrary joint number greater than or equal to i. We investigate the change in H_{ii} when the joint angle θ_l varies from 0 to 360 degrees. As in the case of Figure 4-3, we immobilize all the joints beyond l and treat the last $(n-l+1)$ links as a single rigid body. Again we use the inertial reference from the centroid $C_{l,n}$ to the joint axis $\mathbf{b}_l$. From Equ. (4.14),

$$H_{ii}^{(l)} = \overline{H}_{ii} + \mathbf{b}_i^T \mathbf{A}\mathbf{N}\mathbf{A}^T \mathbf{b}_i + M\left(|\,\mathbf{r}_{i,l} + \mathbf{p}_l\,|^2 - [\mathbf{b}_i^T(\mathbf{r}_{i,l} + \mathbf{p}_l)]^2\right) \tag{4.30}$$

where $\overline{H}_{ii}$ is a constant parameter independent of θ_l, and is given by

$$\overline{H}_{ii} = \sum_{k=i}^{l-1} \left(m_k\left[|\mathbf{r}_{i,ck}|^2 - (\mathbf{b}_i^T\mathbf{r}_{i,ck})^2\right]\right) \tag{4.31}$$

We first derive the conditions that the diagonal element H_{ii} is invariant for an arbitrary θ_l, when the joint axes, $\mathbf{b}_l$ and $\mathbf{b}_i$, are not orthogonal. Then we discuss the case where the axes are orthogonal to each other. Substituting Equ. (4.20) and Equ. (4.21) into Equ. (4.30) yields

$$H_{ii}^{(l)} = a_0 + a_1\cos^2\theta + a_2\cos\theta\sin\theta + a_3\sin^2\theta + a_4\cos\theta + a_5\sin\theta \tag{4.32}$$

where the coefficients $a_0, ..., a_5$ are given by

$$a_0 = \overline{H}_{ii} + M(L^2 + r_x^2 + r_y^2 + (b_x r_x + b_y r_y)^2) + b_z^2 \overline{N}_{zz} \tag{4.33}$$

$$a_1 = ML^2 b_x^2 + b_x^2 \overline{N}_{xx} + b_y^2 \overline{N}_{yy} + 2b_x b_y \overline{N}_{xy} \tag{4.34}$$

$$a_2 = 2(ML^2 b_x b_y + b_x b_y (\overline{N}_{xx} - \overline{N}_{yy}) - (b_x^2 - b_y^2) \overline{N}_{xy}) \tag{4.35}$$

$$a_3 = ML^2 b_y^2 + b_x^2 \overline{N}_{yy} + b_y^2 \overline{N}_{xx} - 2b_x b_y \overline{N}_{xy} \tag{4.36}$$

$$a_4 = 2\left(b_z(b_x \overline{N}_{xz} + b_y \overline{N}_{yz}) + M(b_x L(b_x r_x + b_y r_y) + L r_x)\right) \tag{4.37}$$

$$a_5 = 2\left(b_z(b_y \overline{N}_{xz} + b_x \overline{N}_{yz}) + M(b_y L(b_x r_x + b_y r_y) + L r_y)\right) \tag{4.38}$$

In order for the $H_{ii}^{(l)}$ to be invariant for an arbitrary θ_l, it is necessary that the coefficients a_i satisfy all of the following conditions:

$$a_1 = a_3 \tag{4.39}$$

$$a_2 = 0 \tag{4.40}$$

$$a_4 = 0 \tag{4.41}$$

$$a_5 = 0 \tag{4.42}$$

The above conditions can be further reduced by combining them with Equs. (4.23) and (4.24), which are the necessary conditions for the off-diagonal element H_{il} to be invariant for an arbitrary θ_l. Substituting Equ. (4.23) into Equ. (4.41),

$$(2b_x^2 + b_y^2) r_x + b_x b_y r_y = 0 \tag{4.43}$$

From Equ. (4.24) and Equ. (4.42),

$$(b_x^2 + 2b_y^2)r_y + b_x b_y r_x = 0 \qquad (4.44)$$

Eliminating r_x in the above two equations, we obtain $(b_x^2 + b_y^2)r_y = 0$, which is satisfied if, and only if, one of the following two conditions is met:

i) $b_x = b_y = 0$, namely, joint axes $\mathbf{b}_i$ and $\mathbf{b}_l$ are parallel, or

ii) $r_x = r_y = 0$, namely, the joint axes intersect at the origin O.

In the same way, we can derive a similar expression, $(b_x^2 + b_y^2)^2\overline{N}_{xy} = 0$, from Equ. (4.39) and Equ. (4.40). This leads to the following two conditions:

i) $b_x = b_y = 0$, or

ii) $\overline{N}_{xy} = 0$ and $\overline{N}_{xx} + ML^2 = \overline{N}_{yy}$

Similarly, necessary conditions can be derived for two joints which are orthogonal to each other. Summarizing the above results, the following proposition is obtained.

Proposition (4.4): The necessary condition for the diagonal element H_{ii}, as well as the off-diagonal element H_{il}, to be invariant for an arbitrary θ_l is that the kinematic structure and mass properties of the manipulator arm satisfy one of the following four conditions:

i) $\mathbf{b}_i = \mathbf{b}_l$, i.e., the two joint axes are parallel.

ii) $\mathbf{b}_i^T \mathbf{b}_l \neq 0$ and
$$r_x = r_y = 0, \text{ i.e. } O_i = O, \text{and}$$
$$\overline{N}_{xy} = 0, \text{and}$$
$$\overline{N}_{xx} + ML^2 = \overline{N}_{yy}.$$

iii) $\mathbf{b}_i^T \mathbf{b}_l = 0$ and
$$L = 0 \text{ and}$$
$$\overline{N}_{xy} = 0 \text{ and}$$
$$\overline{N}_{xx} = \overline{N}_{yy}.$$

iv) $\mathbf{b}_i^T \mathbf{b}_l = 0$ and
$$r_y = 0$$
$$\overline{N}_{xy} = 0$$
$$\overline{N}_{xx} = \overline{N}_{yy} + ML^2.$$

4.5 Design

4.5.1 Arm Design for Decoupled Inertia

Based on the previous analysis on the necessary conditions, we determine the kinematic structure and mass properties of a manipulator arm that possesses decoupled and/or invariant inertia matrices.

First, we consider the decoupled inertia. From Proposition (4.2) it follows that all the joint axes must be orthogonal for an arbitrary arm configuration to accomplish the decoupled inertia. The kinematic structure in which all the joint axes are orthogonal at all time is limited to two degree-of-freedom arms only.

For the inertia matrix of the 2 d.o.f. arm, the off-diagonal element to be considered is only H_{12}, and the joint desplacement that might cause changes in H_{12} is θ_2. Then we can derive the necessary conditions for invariant inertia by applying Proposition (4.3) only to $H_{12}^{(2)}$.

Conversely, if the necessary conditions on $H_{12}^{(2)}$ are substituted into Equ. (4.22) to compute the off-diagonal element H_{12}, one can find that the off-diagonal element vanishes for an arbitrary θ_2. Consequently, the necessary conditions also provide the sufficient conditions for decoupled inertia. Therefore the following obtained:

> **Theorem (4.1):** An open-kinematic-chain manipulator arm possesses a decoupled inertia matrix for an arbitrary arm configuration if, and only if, its kinematic structure and mass properties satisfy the following conditions:
> Structure: two degree-of-freedom arm with orthogonal joint axes (Figure 4-5).

Mass properties: $m_2 L r_z = \overline{N}_{xz}$ (4.45)

and

$$\overline{N}_{yz} = 0, \tag{4.46}$$

where the notations are defined in Figure 4-5.

Special cases, which are practically useful, are shown in Figure 4-6. In the first arm design, the centroid of link 2 lies on the joint axis b_2, namely, $L = 0$, and in the second arm design the perpendicular from the centroid of link 2 to the joint axis b_2 intersects with the common normal of the two joint axes, namely, $r_z = 0$. For each of the two cases, the left hand side of Equ. (4.45) is zero, hence $\overline{N}_{xz}$ must be coincident with the joint axis b_2, since $\overline{N}_{yz}$ is also zero from condition Equ. (4.46).

4.5.2 Arm Design for Decoupled and Invariant Inertia

Manipulator arms with decoupled and configuration-invariant inertia matrices are discussed in this subsection. To be decoupled, the conditions given in Theorem (4.1) must hold. At the same time, the diagonal element H_{11} must

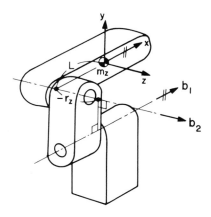

Figure 4-5: A 2 d.o.f. open kinematic chain manipulator with decoupled inertia tensor.

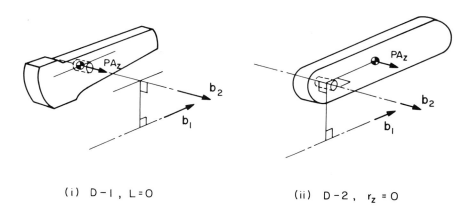

(i) D – I , L = O   (ii) D – 2 , r_z = O

Figure 4-6: Designs for 2 d.o.f. manipulators with decoupled inertia tensor.

be constant to meet the requirement for the configuration-invariant inertia. It is therefore necessary that the conditions in Proposition (4.4) are satisfied for the diagonal element. The first two cases involved in Proposition (4.4) conflict with the decoupling conditions, because the manipulator arm that can possess a decoupled inertia matrix is limited to one with orthogonal joint axes. Thus, the set of conditions, either (iii) or (iv), must be met. Conversely, by substituting either of conditions (iii) or (iv) of Proposition (4.4) as well as the conditions in Theorem (4.1), into Equ. (4.22) and Equ. (4.30), one can find that the (2×2) inertia matrix becomes diagonal and constant, hence these conditions are sufficient conditions as well.

Theorem (4.2): The necessary and sufficient conditions for an open-kinematic-chain manipulator arm to possess a decoupled and configuration-invariant inertia matrix are given by:

Structure: two degree-of-freedom arm with orthogonal joint axes,

Mass properties: one of the following two sets of conditions must hold:

i)
$$L = 0 \text{ and} \tag{4.47}$$

$$\overline{N}_{xx} = \overline{N}_{yy} \text{ and} \tag{4.48}$$

$$\overline{N}_{xy} = \overline{N}_{yz} = \overline{N}_{xz} = 0 \tag{4.49}$$

ii)
$$r_y = 0 \text{ and} \tag{4.50}$$

$$\overline{N}_{xx} = \overline{N}_{yy} + ML^2 \text{ and} \tag{4.51}$$

$$\overline{N}_{xy} = \overline{N}_{yz} = 0 \text{ and} \tag{4.52}$$

$$m_2 L r_z = \overline{N}_{xz} \tag{4.53}$$

The arm designs corresponding to the above two cases are shown in Figure 4-7.

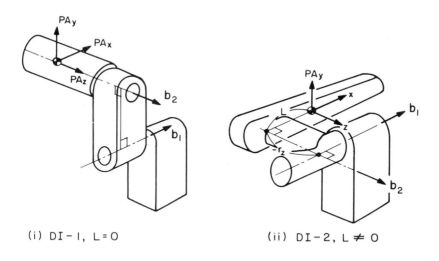

(i) DI – I, L = O   (ii) DI – 2, L ≠ O

Figure 4-7:   Designs for 2 d.o.f. manipulators with decoupled and
                  configuration invariant inertia tensors.

For convenience, the two arms are called model DI-1 and model DI-2, respectively. From Equ.(4.49), the second link of model DI-1 has its principal directions in the same directions as joint axes b_1 and b_2. The direction perpendicular to both b_1 and b_2 is then the principal direction as well. For model DI-2, on the other hand, one of the principal directions of link 2 is along the PA_y axis shown in the figure.

4.5.3 Arm Design for Invariant Inertia

In this subsection we discuss the arm design when only the configuration invariance of the inertia matrix is required. Since the orthogonality condition is not needed in this case, the arm design can be extended to more than 2 degrees of freedom. However, we first discuss 2 d.o.f. arms.

4.5.3.1 Two degree-of-freedom arms

We investigate the elements H_{11} and H_{12} in relation to θ_2. Conditions involved in Propositions (4.1), (4.3), and (4.4) are now listed in Table 4-1, from which conditions for the invariant inertia are to be derived. The upper half of the table is for the conditions on $H_{12}^{(2)}$, while the lower half is for $H_{11}^{(2)}$. Since both elements must be constant for all θ_2, the upper and lower sets of conditions must be satisfied at the same time. For each of the conditions, either the upper quarter of the table or the lower one must hold, thus the two sets of conditions are led to the "OR" gates representing that either one set or the other must be satisfied. These conditions can be reduced to the following.

When $b_z = 0$, the two joint axes are orthogonal. Therefore the conditions for the invariant inertia are the same as in Theorem (4.2). We then focus on the case where $b_z \neq 0$. From the table, two cases arise; (i) one is the set of conditions 1, 2, 3, and 8, (ii) the other is conditions 1, 2, 3, 9, 10, 11, and 12. In case (i), substituting condition 8 as well as its equivalent conditions, $b_x = b_y = 0$, into conditions 2 and 3, we find that $Lr_x = 0$ and $Lr_y = 0$. If we assume that $L \neq 0$, then r_x and r_y must be zero. What this means is that the two joint axes are identical, which is a trivial case, hence neglected. Thus the distance L must be zero for case (i). In case (ii), substituting conditions 10 and 11 into 2 and 3 yields $(b_x^2 + b_y^2)\overline{N}_{yz} = 0$. Since b_x and b_y are not zero at the same time, $\overline{N}_{yz}$ must be zero. In the same way, we can derive that $\overline{N}_{xz} = 0$. From these results along with condition 9, it follows that the three principal directions of link 2 are, respectively, along joint axis b_2, the perpendicular from the centroid C_2 to the

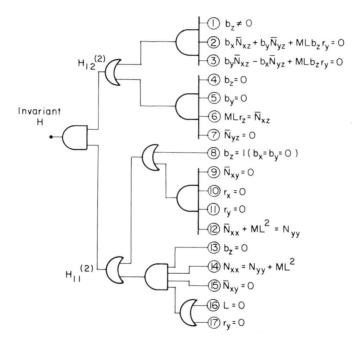

Table 4-1: Necessary conditions for 2 d.o.f. arm to possess invariant inertia matrix

joint axis b_2, and the perpendicular to both axes. The above investigations are summarized in the following theorem.

Theorem (4.3): The two degree-of-freedom manipulator arm consisting of an open kinematic chain possesses a configuration-invariant inertia matrix if, and only if, one of the following four conditions is satisfied:

i)
$$b_z = 1, \quad (b_x = b_y = 0) \text{ and} \tag{4.54}$$

$$L = 0 \tag{4.55}$$

ii)
$$r_x = r_y = 0 \tag{4.56}$$

$$\overline{N}_{xy} = \overline{N}_{yz} = \overline{N}_{xz} = 0 \tag{4.57}$$

$$\overline{N}_{xx} + ML^2 = \overline{N}_{yy} \tag{4.58}$$

iii) the same as condition (i) in Theorem (4.2)

iv) the same as condition (ii) in Theorem (4.2)

The manipulator arms corresponding to the first two cases (i) and (ii) are shown in Figure 4-8. These arm designs are called model I-1 and model I-2, respectively.

4.5.3.2 Three degree-of-freedom arms

Applying the above theorem, we extend the arm design to three degrees of freedom. A 3 d.o.f. arm reduces to a 2 d.o.f. arm when one of the three joints is immobilized. According to Theorem (4.3), the possible designs for 2 d.o.f. arms are limited to the four cases to accomplish the configuration-invariant inertia. Among the four models, model I-2, which includes an oblique joint, is seldom used in practice. We therefore do not adopt this model to our arm design. We build the 3 d.o.f. arm with the combination of the three models only. Table 4-2 gives the list of possible combinations of the three models.

For 3 d.o.f. arms, the inertia matrix must be invariant for an arbitrary combination of joint angles, θ_2 and θ_3 . We investigate whether or not the

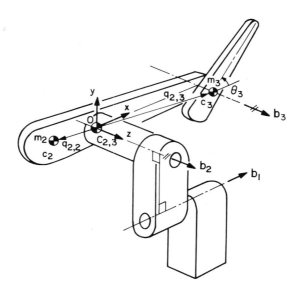

Figure 4-9: Design of a 3 d.o.f. manipulator with a configuration
invariant inertia tensor.

$$N_{2,3} = I_2 + m_2[\,diag\,(q_{2,2}^T q_{2,2}) - q_{2,2}q_{2,2}^T] + A(\theta_3)I_3 A(\theta_3)^T \qquad (4.59)$$
$$+ m_3[\,diag\,(q_{2,3}^T q_{2,3}) - q_{2,3}q_{2,3}^T]$$

On the other hand, the centroid of link 3 must lie on the joint axis b_3, since the last two joints are model I-1. Thus

$$L_{2,3} = 0 \qquad\qquad (4.60)$$

Therefore, position vector $q_{2,3}$ is independent of θ_3. Similarly, the centroid of the composite body, $C_{2,3}$, must lie on the joint axis b_2, because the first two joints are the DI-1 type. Thus,

$$m_2 q_{2,2} + m_3 q_{2,3} = 0 \qquad\qquad (4.61)$$

In addition, according to Equ. (4.49), which is a mass property condition for model DI-1, all the off-diagonal elements of the composite inertia tensor $N_{2,3}$ must be zero for an arbitrary θ_3. For this to be true, the right hand side of Equ. (4.59) is at least independent of θ_3. Since all the terms except the third term in Equ. (4.59) are independent of θ_3, we need to evaluate only the third term. This yields additional conditions for the inertia tensor of link 3 given by

$$I_{3xy} = I_{3yz} = I_{3xz} = 0 \qquad\qquad (4.62)$$

$$I_{3xx} = I_{3yy} \qquad\qquad (4.63)$$

Substituting Equ. (4.60), Equ. (4.61) and Equ. (4.62) into Equ. (4.59) and evaluating the composite inertia tensor $N_{2,3}$, we can derive additional conditions on link 2 that are necessary to satisfy Equ. (4.48) and Equ. (4.49) for an arbitrary θ_3.

$$I_{2xy} = q_x q_y m_2 \left(1 + \frac{m_2}{m_3} \right) \tag{4.64}$$

$$I_{2yz} = q_y q_z m_2 \left(1 + \frac{m_2}{m_3} \right) \tag{4.65}$$

$$I_{2xz} = q_x q_z m_2 \left(1 + \frac{m_2}{m_3} \right) \tag{4.66}$$

$$I_{2xx} = I_{2yy} + (q_x^2 - q_y^2) m_2 \left(1 + \frac{m_2}{m_3} \right) \tag{4.67}$$

As before, the substitution of the above equations into the general inertia matrix in Equ. (4.9) reveals that the above necessary conditions provide the sufficient conditions as well. Thus the set of conditions we derive are necessary and sufficient for the 3 d.o.f. arm consisting of the DI-1 and I-1 joint types. Similarly the other arm designs are analyzed by using Equ. (4.59), and additional mass property conditions are determined. Thus the possible designs for 3 d.o.f. arms with configuration-invariant inertia are all found.

4.6 Discussions on the Decoupling Conditions

In this section, we discuss the necessary conditions derived previously in order to find the design guidelines for decoupling serial spatial mechanisms. The first requirement in the necessary conditions is the orthogonality of the joint axes. In all cases studied during the derivation of the necessary conditions, the orthogonality of joint axes was required to eliminate the torque due to the centroidal rotation. This is the torque due to the rotational motion of a link about its center of mass. Another torque results from the motion of the center of mass. This motion is referred to as the motion of the centroid. These terms were defined in references [Youcef- Toumi and Asada 1985] and [Youcef-Toumi 1985]. We can conclude that the coupling between the joints of a serial spatial mechanism is dictated by the centroidal rotations of the links.

The reason for this is the following. The reaction torques due to the motion of the centroid can be cancelled using mass distribution or modifying the arm structure as the necessary conditions suggest. However, the rotational effects are cancelled only by having orthogonal joint axes, as dictated explicitly by the expression giving the reaction torques

$$\tau = [I+mk^2]b_i b_j \ddot{\theta}.$$

where

$$k^2 \geq 0.$$

Therefore, the motions due to the centroidal rotation limit the decoupling of serial spatial mechanisms by mass distribution only, and we must resort to an additional tool in order to decouple more than two d.o.f.

Since the centroidal rotation of links is the limiting factor, it can be stated, from the mass properties point of view, that the moments of inertia characterize the decoupling by mass distribution. Thus in serial drive mechanisms, in order to achieve decoupling, the rotational effects must be eliminated. The orthogonality of joint axes as given by the necessary condition, does eliminate the effects of pure rotation of the links. Therefore, any other means that can eliminate the effects of centroidal rotation will be equivalent to the orthogonality condition. The following corollary can be stated

> **Corollary (4.1):** Decoupling of the manipulator inertia tensor for more than two degrees of freedom cannot be accomplished merely by redistributing mass, the structure of the arm mechanism itself needs to be modified.

In addition, the orthogonality condition required by the necessary condition cannot be satisfied for serial planar mechanisms where all joint axes are parallel, thus a corollary to the previous theorems is

> **Corollary (4.2):** It is impossible to decouple the inertia tensor of planar open loop kinematic chains that use revolute joints with actuators mounted at joints and exert torques between adjacent links.

4.7 Conclusion

This chapter first introduced the concepts of decoupled and/or configuration-invariant inertia. The design theory for accomplishing those particular forms was presented. In addition, the necessary conditions pertaining to the kinematic structure and mass distributions for arms to possess decoupled and/or configuration-invariant inertia were derived. These necessary conditions provide useful guidelines for the arm linkage design.

For decoupling, all the joint axes must be orthogonal. Consequently, manipulator arms with more than two degrees of freedom cannot be dynamically decoupled by design. Based on this design guideline, all of the arm constructions that yielded the decoupled inertia matrices were determined.

Using the necessary conditions, the effects due to centroidal rotation of the links were shown to be the limiting factor in the decoupling of manipulator inertia tensors by mass distribution. It was also pointed out that decoupling of more than two degrees of freedom can be achieved by a means that eliminates the effects of centroidal rotation but does not use joint axes orthogonality. This will be discussed in detail in the next chapter.

For the configuration-invariant inertia, there is no limitation regarding the number of degrees of freedom. In this chapter, 2 and 3 degree-of-freedom arms were considered. For more than three degrees of freedom, the same approach can be applied. However, the higher the number of d.o.f., the more difficult finding practically useful arm designs becomes.

Chapter 5

Actuator Relocation

5.1 Introduction

The approach presented in this previous Chapter is to simplify the arm dynamics by design. Using adequate mass distributions for the arm links a form of the manipulator inertia tensor for which the coupling is significantly reduced can be obtained. Consequently, the manipulator dynamics is simplified and the computation burden is relaxed. This design technique is related to force balancing. Force balance deals with the redistribution of mass to reduce the shaking forces transmitted from the mechanism to the frame or ground. Most of this research has been directed to planar mechanisms [Tepper and Lowen 72, Tepper and Lowen 73, Tepper and Lowen 75, Walker and Oldham 78]. In our context, the mass redistribution are used to modify the dynamics of multi-degree-of-freedom linkage system with multiple actuators. Specifically, the inertia tensor of the linkage system is to be made decoupled and invariant. This decoupled and invariant dynamic behavior is particularly desirable in manipulators performing high-speed and high-accuracy path control tasks.

The feasibility and practical usefulness have been demonstrated through the development of special direct-drive arms. Analysis and procedure in design of manipulator arms with open loop kinematic chains was made in a comprehensive form in the previous chapter and in references [Asada and Youcef-Toumi 84, Asada and Youcef-Toumi 84, Youcef-Toumi 85, Youcef-Toumi and Asada 85a, Youcef-Toumi and Asada 85b, Youcef-Toumi and Asada 85, Youcef-Toumi and Asada 86]. In references [Yang and Tzeng 85, Yang and Tzeng 86], a different approach was adopted. Namely, the total kinetic and potential energy are calculated, then the designer examines all the terms involved in the energy expression to decide what mass distributions should be selected to achieve simple dynamics for the given manipulator. In the following chapters, the conditions for the kinematic structure and mass properties of the manipulator arm which provide a decoupled and/or configuration-invariant inertia for planar and spatial mechanisms are determined. Specifically, the interest is to derive necessary and sufficient conditions that must satisfied to achieve such desirable dynamic behavior.

In the previous Chapter, we assumed that each actuator was located at the corresponding joint and that it exerts a torque between the adjacent links. The actuator, however, can be located on other links if an appropriate transmission can be used. In this Chapter, we discuss the decoupling problem for a mechanism where the actuators are relocated to other links with the use of a transmission. It will be shown that the decoupling condition, previously obtained, can be significantly relaxed by the relocation of actuators [Youcef-Toumi 85, Youcef-Toumi and Asada 85b]. When actuators are mounted remotely and joints are driven through a transmission, the torque exerted by one actuator is not necessarily applied to one joint but can affect several joints. Similarly, the displacement of one actuator may cause a displacement in more than one joint axis. Thus, we need to study the effects of relocating the actuators on the manipulator kinematics and dynamics. First, we set up the relationship between the actuator displacements and their corresponding joint displacements . Then we determine the linear and angular velocity vectors of every link in order to examine the new manipulator dynamics. Second, we focus on the manipulator mass properties. We recall that the effects of the centroidal rotation of links were the limiting factor in decoupling serial drive manipulators. Therefore, it is important to determine which link mass properties effect the coupling between any two given actuators. Specifically, the moments of inertia are of special interest since they, in effect, limit the decoupling. A useful set notation is introduced to facilitate the analysis and help in the understanding of coupling.

5.2 Modeling of Transmission

The class of manipulators examined consists of serial linkages with all revolute joints. We assume that the actuator displacements $(\alpha_1, \alpha_2, ..., \alpha_n)$ are a complete set of independent generalized coordinates that are able to locate the manipulator uniquely and completely. This section presents a model of the transmission for this class of manipulators and introduces the following notations,

   • α_l : the l-th actuator displacement

- θ_m : the mth joint displacement defined according to the Denavit and Hartenberg notation.

- b_m : unit vector in the direction of the m-th joint axis

Thus, the lth actuator produces an actuator displacement α_l which induces link m to rotate, through a transmission, by a displacement θ_m about joint axis b_m . In the analysis, we assume that the inertia of the transmission is negligible and we deal only with transmission of static forces. The actuator displacement vector, α, and the joint displacement vector, Θ, are assumed to be related by the vector function f^1,

$$\Theta = f\{\alpha\} \tag{5.1}$$

and the actuator velocities α and joint velocities Θ are related by

$$\dot{\Theta} = K(\alpha)\dot{\alpha} \tag{5.2}$$

where $K(\alpha)$ is the Jacobian matrix associated with f, and depends on the arm configuration in general. We will refer to the matrix K as the transmission matrix. For a serial arm with actuators mounted at the joints between adjacent links, the transmission matrix K is a diagonal constant matrix with diagonal elements equal to the inverse of the gear ratios. Figures 5-1 and 5-2 show two two-degree-of-freedom manipulators, where the actuators are mounted at the joints and at the base, respectively. As shown in the Figures, the transmission matrices, K, depend on the location of the actuators. In what follows, we derive the expression for the angular and the linear velocity vectors.

5.2.1 Link Angular Velocity Vector

In general, the angular velocity of link i, w_i is given by

$$w_i = \sum_{l=1}^{n} \left(\partial w_i / \partial \dot{\alpha}_l \right) \dot{\alpha}_l \tag{5.3}$$

---

[1]It may be more appropriate to write $\alpha = f(\theta)$ when the dim $(\theta) >$ dim (α).

Direct-Drive Robots

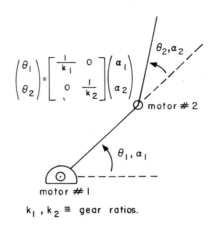

Figure 5-1: A 2 d.o.f serial manipulator with motors at the joints

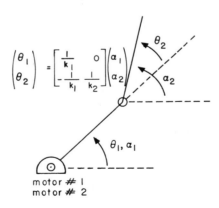

Figure 5-2: A 2 d.o.f serial manipulator with motors at the base

where α_l is the velocity of actuator l, and $\partial \mathbf{w}_i / \partial \dot{\alpha}_l$ is Kane's partial rate of change of orientation with respect to α_l [Kane 68]. This partial rate is denoted by $\mathbf{a}_{il}$. In the case, where $\partial \mathbf{w}_i / \partial \dot{\alpha}_l$ is zero, actuator l induces no angular velocity to link i. Thus, we can define a set S_i of all actuators j that induce rotation to link i

$$S_i = \left\{ j \,\middle|\, \frac{\partial \mathbf{w}_i}{\partial \alpha_j} \neq 0 \right\} \tag{5.4}$$

and

$$\mathbf{w}_i = \sum_{l \in S_i} \left(\partial \mathbf{w}_i / \partial \dot{\alpha}_l \right) \dot{\alpha}_l$$

For a serial manipulator, the velocity vector $\mathbf{w}_i$ is given by

$$\mathbf{w}_i = \sum_{j=1}^{i} \mathbf{b}_j(\theta) \dot{\theta}_j$$

Equ. (5.2), rewritten as

$$\dot{\theta}_j = \sum_{l=1}^{n} K_{jl} \dot{\alpha}_l \tag{5.5}$$

therefore becomes

$$\mathbf{w}_i = \sum_{j=1}^{i} \mathbf{b}_j \sum_{l=1}^{n} K_{jl} \dot{\alpha}_l = \sum_{l=1}^{n} \left(\sum_{j=1}^{i} \mathbf{b}_j K_{jl} \right) \dot{\alpha}_l = \sum_{l \in S_i} \mathbf{a}_{il} \dot{\alpha}_l \tag{5.6}$$

Now, we can write explicitly the expression for the partial rate of orientation $\mathbf{a}_{il}$ with respect to α_l,

$$\left(\partial \mathbf{w}_i / \partial \dot{\alpha}_l \right) = \mathbf{a}_{il} = \sum_{j=1}^{i} \mathbf{b}_j K_{jl} \tag{5.7}$$

Thus, in this case, if it is desired that an actuator l is to have no contribution to the rotation of link i, $\mathbf{a}_{il}$ must be zero. This means that the linear combination of the unit vectors, $\mathbf{b}_j$, weighted by the the elements of the transmission matrix, $\mathbf{K}$, must be zero. If the vectors $\mathbf{b}_j$ are linearly dependent, we need to choose some nonzero K_{jl}'s so that Equ. (5.7) vanishes.

The definition of S_i for a serial manipulator becomes

$$S_{i_{serial}} = \left\{ l \mid \sum_{j=1}^{i} \mathbf{b}_j K_{jl} \neq 0 \right\} \tag{5.8}$$

5.2.2 Link Linear Velocity Vector

The linear velocity vector $\mathbf{v}_i$ of the mass center (ic) of link i, is

$$\mathbf{v}_i = \sum_{l=1}^{n} \left(\partial \mathbf{v}_i / \partial \dot{\alpha}_l \right) \dot{\alpha}_l \tag{5.9}$$

where $\partial \mathbf{v}_i / \partial \dot{\alpha}_l$ is Kane's partial rate of change of position with respect to α_l. This partial rate is denoted by $\mathbf{a}_{il}^*$. We define a new set S_i^*, similar to S_i, which is the set of all actuators that induce linear motion to the mass center of link i,

$$S_i^* = \left\{ j \mid \left(\partial \mathbf{v}_i / \partial \dot{\alpha}_j \right) \neq 0 \right\} \tag{5.10}$$

and

$$\mathbf{v}_i = \sum_{l \in S_i^*} \left(\partial \mathbf{v}_i / \partial \dot{\alpha}_l \right) \dot{\alpha}_l$$

For a serial manipulator, the velocity vector $\mathbf{v}_i$ is

$$\mathbf{v}_i = \sum_{j=1}^{i} \mathbf{b}_j \times \mathbf{r}_{ic}^j \, \dot{\theta}_j \qquad (5.11)$$

where $\mathbf{r}_{ic}^j$ is a vector from joint axis j to the mass center of link i. Now using Equ. (5.5), we have

$$\mathbf{v}_i = \sum_{j=1}^{i} \mathbf{b}_j \times \mathbf{r}_{ic}^j \left(\sum_{l=1}^{n} K_{jl} \dot{\alpha}_l \right)$$

or

$$\mathbf{v}_i = \sum_{l=1}^{n} \left(\sum_{j=1}^{i} K_{jl} \mathbf{b}_j \times \mathbf{r}_{ic}^j \right) \dot{\alpha}_l \qquad (5.12)$$

By inspection, we recognize that the partial rate of change of position with respect to α_l is

$$\mathbf{a}_{il}^* = \partial \mathbf{v}_i / \partial \dot{\alpha}_l = \sum_{j=1}^{i} K_{jl} \mathbf{b}_j \times \mathbf{r}_{ic}^j$$

And the set, S_i^*, for a serial manipulator becomes

$$S_{i_{serial}}^* = \left\{ l \,\middle|\, \sum_{j=1}^{i} K_{jl} \mathbf{b}_j \times \mathbf{r}_{ic}^j \neq 0 \right\} \qquad (5.13)$$

for all arm configurations.

5.3 Kinetic Energy

5.3.1 Arm Inertia Tensor

In this section we examine the total kinetic energy KE of an n degree-of-freedom mechanism with n links, in terms of the actuator displacements, in order to evaluate the effect of the transmission on the elements of the manipulator inertia tensor. The kinetic energy in terms of the actuator displacements α is

$$KE = \sum_{i=1}^{n} KE_i = \frac{1}{2} \sum_{i=1}^{n} (m_i \mathbf{v}_i^T \mathbf{v}_i + \mathbf{w}_i^T \mathbf{I}_i \mathbf{w}_i)$$

(5.14)

$$KE = \frac{1}{2} \dot{\alpha}^T \mathbf{G} \dot{\alpha}$$

where $\mathbf{v}_i$ and $\mathbf{w}_i$ are the linear and angular velocities of link i expressed in terms of the actuator velocities, and m_i and $\mathbf{I}_i$ are the ith link mass and inertia tensor respectively. The kinetic energy can be rewritten in terms of the actuator velocities explicitly; matrix $\mathbf{G}$ is the inertia tensor for this expression. Let us derive $\mathbf{G}$. Substituting the general expressions of the angular and linear velocities given by Equ. (5.3) and Equ. (5.9), in terms of the partial rates $\mathbf{a}_{il}$ and $\mathbf{a}_{il}^*$, the kinetic energy is

$$KE = \frac{1}{2} \sum_{i=1}^{n} \left[m_i \left(\sum_{l \in S_i^*} \dot{\mathbf{a}}_{il}^* \dot{\alpha}_l \right)^T \left(\sum_{m \in S_i^*} \mathbf{a}_{im}^* \dot{\alpha}_m \right) \right.$$

(5.15)

$$\left. + \left(\sum_{l \in S_i} \mathbf{a}_{il} \dot{\alpha}_l \right)^T \mathbf{I}_i \left(\sum_{m \in S_i} \mathbf{a}_{im} \dot{\alpha}_m \right) \right]$$

and the $l{-}m$ element, h_{lm} of the inertia tensor $\mathbf{G}$ can be identified as

$$h_{lm} = \sum_{i \in S_{lm}} (\mathbf{a}_{il})^T \mathbf{I}_i (\mathbf{a}_{im}) + \sum_{i \in S_{lm}^*} m_i (\mathbf{a}_{il}^*)^T (\mathbf{a}_{im}^*)$$

(5.16)

where the sets S_{lm} and S^*_{lm} are defined as

$$S_{lm} = \left\{ i \,\middle|\, l \in S_i \text{ and } m \in S_i \right\}$$

$$(5.17)$$

$$S_{lm}^* = \left\{ i \,\middle|\, l \in S_i^* \text{ and } m \in S_i^* \right\}$$

The set, S_{lm}, is the set of all links i that are rotated by both actuators l and m. On the other hand, the set S_{lm}^* is the set of all links i whose centroids motions are induced by both actuators l and m. These sets explicitly show how the mass properties of the different links are distributed in the manipulator inertia tensor. The mass properties of these links will thus appear in the $l–m$ element of the manipulator inertia tensor (without regard to cancelation of terms). If, on the other hand, link i does not belong to either set S_{lm} or S_{lm}^*, then its mass properties are not involved in the $l–m$ element of the manipulator inertia tensor.

Equation (5.16) is a general expression for the $l–m$ element of the arm inertia tensor **G**. First we note that the elements of the transmission matrix **K**, namely K_{il}, are involved in this expression. They are additional design parameters that could be used to make the off-diagonal elements h_{lm}, $l \neq m$, zero. The first term in Equ. (5.16), is related to the reaction torque resulting from the centroidal rotation. Similarly, the second term is related to the motion of the link's centroid. Using the proposition given in Chapter 4, we can state that the first term in h_{lm} must be eliminated by some means other than the orthogonality condition.

In general, the links whose mass properties would definitely have no contributions[2] to the value of h_{lm} are those links i for which the partial rates, corresponding to actuators l and m, are zero. Specifically, the mass (moments of inertia) of link i would not appear in h_{lm} when the partial rate $\mathbf{a}^*_{il}$ or $\mathbf{a}^*_{im}$ ($\mathbf{a}_{il}$ or $\mathbf{a}_{im}$) is zero for all arm configurations. In other words, actuator l or m does not induce motion (rotation) to (about) the mass center of link i;

---

[2]Other possibilities for which a link's mass properties do not contribute to the value of h_{lm} are

- $(\mathbf{a}^*_{il})^T \mathbf{a}^*_{im} = 0$, that is orthogonal vectors.
- $(\mathbf{a}_{il})^T \mathbf{I}_i \, \mathbf{a}_{im} = 0$, that is $\mathbf{a}_{il}$ and $\mathbf{a}_{im}$ are $\mathbf{I}_i$ orthogonal.

$l \notin S_i^*$ or $m \notin S_i^*$ that is $\mathbf{a}_{il}^* = 0$ or $\quad \mathbf{a}_{im}^* = 0$

$l \notin S_i$ or $m \notin S_i$ that is $\mathbf{a}_{il} = 0$ or $\mathbf{a}_{im} = 0$

5.3.2 Examples

Example 1:

Consider the manipulator shown in Figure 5-2 and assume that the drive systems have unity gear ratios. The influence of the velocity of actuator 1, $\dot{\alpha}_1$, on the angular velocity vector $\mathbf{w}_2$ of link 2 is

$$\mathbf{a}_{21} = \frac{\partial \mathbf{w}_2}{\partial \dot{\alpha}_1} = \mathbf{b}_1 K_{11} + \mathbf{b}_2 K_{21}$$

where $\mathbf{b}_1$ and $\mathbf{b}_2$ are parallel joints axes. The elements of the transmission matrix, K_{11} and K_{21}, are indentified as 1 and -1 respectively, as shown in Figure 5-2. Thus, one obtains

$$\frac{\partial \mathbf{w}_2}{\partial \dot{\alpha}_1} = \mathbf{b}_1(1) + \mathbf{b}_2(-1) = 0$$

Therefore, $\dot{\alpha}_1$ has no influence on $\mathbf{w}_2$ since the partial rate of change of orientation of link 2, with respect to α_1, is zero. In fact, the angular velocity vector $\mathbf{w}_2$ is

$$\mathbf{w}_2 = \mathbf{b}_2 \dot{\alpha}_2$$

so that the set S_2 contains motor 2 only,

$$S_2 = \{2\}$$

Example 2:

Both manipulators of Figures 5-1 and 5-2 have an S^*_2 given by

$$S^*_2 = \{1,2\}$$

This means that the motion of the mass center of link 2 in both manipulators is induced by both motors 1 and 2.

Example 3:

If we apply the above procedure to the manipulator of Figure 5-2, we obtain $S_1 = \{1\}$ and $S_2 = \{2\}$, since motor 1 rotates link 1 only and motor 2 rotates link 2 only.

$S^*_1 = \{1\}$ and $S^*_2 = \{1,2\}$, motion of mass center of link 1 is induced by motor 1 only, and that of link 2 is induced by both actuators 1 and 2.

Thus, using the definition given by Equ. (5.17), the sets S_{lm} and S^*_{lm} are:

$S_{11} = \{1\}$, $S_{12} = \varnothing$, $S_{22} = \{2\}$, $S^*_{11} = \{1,2\}$, $S^*_{12} = \{2\}$, and $S^*_{22} = \{2\}$

Therefore, the elements of the inertia tensor, as given by Equ. (5.16) should involve the following mass properties

$$h_{11} = h_{11}(m_1, m_2, I_1)$$

$$h_{12} = h_{12}(m_2)$$

$$h_{22} = h_{22}(m_2, I_2)$$

Example 4:

Let us analyze the link mass properties of the parallel drive five-bar-link mechanism of Figure 3-6

Link Rotations: $S_1 = \{1\}$, $S_2 = \{2\}$, $S_3 = \{1\}$, $S_4 = \{2\}$

Centroids motions: $S^*_1 = \{1\}$, $S^*_2 = \{2\}$, $S^*_3 = \{1,2\}$, $S^*_4 = \{1,2\}$

Using these descriptions, the S_{lm} and S^*_{lm} sets are found to be:

$$S_{11} = \{1,3\} \text{ and } S_{11}^* = \{1,2,3\}$$

$$S_{22} = \{2,4\} \text{ and } S_{22}^* = \{2,3,4\}$$

$$S_{12} = \varnothing \text{ and } S_{12}^* = \{3,4\}$$

thus the functional relationship for the elements of the inertia tensor are as follows:

$$h_{11} = h_{11}(m_1, m_3, m_4, I_1, I_3)$$

$$h_{22} = h_{22}(m_2, m_3, m_4, I_2, I_4)$$

$$h_{12} = h_{12}(m_3, m_4)$$

Note that in this case, the off-diagonal element h_{12} results only from the linear motion of links 3 and 4. The general expressions of the inertia tensor are

$$h_{11} = I_1 + I_3 + m_1 g^2{}_1 + m_3 g^2{}_3 + m_4 l^2{}_1$$

$$h_{22} = I_2 + I_4 + m_2 g^2{}_2 + m_3 l^2{}_2 + m_4 g^2{}_4$$

$$h_{12} = (m_3 l_2 g_3 - m_4 l_1 g_4) \cos(\theta_2 - \theta_1)$$

5.3.3 Transmission Matrices for Serial Manipulators

In this section we derive the general form of the transmission matrices for open loop revolute chains. The actuators can be mounted anywhere at the joints between the adjacent links. A typical transmission is shown in Figure 5-3, where motor i is mounted on link k, and drives link i. The joint displacements that are involved between the ith motor and the joint about which link i rotates range from $k+1$ to $i-1$. Following the Denavit and Hartenberg notation, that is link i rotates about joint axis $i-1$, we assign a value of zero to k to represent ground. In other words, link 1 rotates about joint axis indexed $k=0$. Now, the differential displacement $\delta\theta_i$ of link i is equal to the sum of the joint displacements ranging from $k+1$ to $i-1$, subtracted from the ith motor displacement $\delta\alpha_i$,

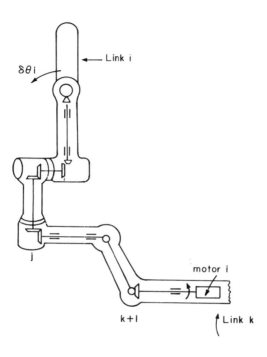

Figure 5-3:   Transmission mechanism in a serial manipulator

$$\delta\theta_i = \delta\alpha_i - \sum_{j=k+1}^{i-1} \delta\theta_j \qquad (5.18)$$

which can be written as

$$\delta\alpha_i = \sum_{j=k+1}^{i} \delta\theta_j \qquad (5.19)$$

5.4 Generalized Forces

In section 5.2 of this Chapter, we introduced a relationship between rates of the generalized joint displacements Θ and the generalized actuator displacements α, namely

$$\dot{\Theta} = \mathbf{K}(\alpha)\dot{\alpha}$$

From the energy equation, the generalized joint forces τ and actuator forces $\mathbf{Q}$ should be related by the transpose of the transmission matrix $\mathbf{K}$,

$$\mathbf{Q} = \mathbf{K}^T \tau$$

5.5 Conclusion

This Chapter has focused on the kinematics and dynamics of mechanisms with transmissions. In particular, it emphasised the effects of actuator locations on the dynamics. Furthermore, a notation using sets was introduced to help understanding the dynamics. In this notation, link motions are represented by the sets S_i and S_i^*, and the link contributions to element h_{lm} of the inertia tensor by the sets S_{lm} and S_{lm}^*. These sets will be used later to set up conditions for decoupling and invariance of manipulator dynamics.

Chapter 6

Design of Decoupled Arm Structures

6.1 Introduction

This Chapter provides some decoupling design conditions which are based on the results of the theorems and corollaries of Chapter 4 and the set notation introduced in Chapter 5. Specifically, the centroidal rotation torque condition is expressed as a design rule for decoupling spatial serial mechanisms. This is then used in deriving the sufficient condition for a decoupled and invariant manipulator inertia tensor.

6.2 Decoupled and Invariant Inertia Tensors for Planar Mechanisms

6.2.1 Necessary Conditions for a Decoupled Inertia Tensor

Consider a planar open loop kinematic chain consisting of n links and driven by revolute actuators mounted at serial joints and exerting torques between adjacent links. In this section, we derive the necessary condition that must be satisfied in order to decouple the inertia tensor of such planar manipulators as shown in Figure 6-1.

Consider the case where all joints displacements $\theta_1, \ldots, \theta_{n-1}$ are immobilized, and only link n is accelerating. In this case, the reaction torque created by the motion of link n and transmitted to link $n-1$ is

$$\tau_{react} = [I_n + m_n l_{n-1} g_n \cos \theta_n] \ddot{\theta}_n$$

For this situation, the expression between bracket represents the off- diagonal element $h_{n\,n-1}$ of the inertia tensor. This expression varies with arm configuration depending on θ_n. From decoupling point of view, this expression may vanish for a limited number of configurations specified by θ_n. Further, it is

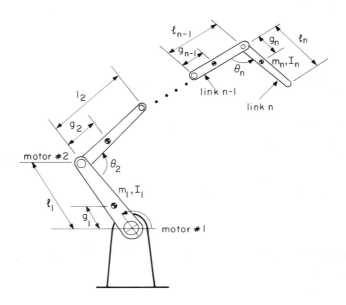

Figure 6-1: An *n* d.o.f. planar manipulator

impossible to choose arm characteristics I_n, m_n, l_{n-1}, and g_n so that h_{nn-1} vanishes for all θ_n. The limiting factor here is the constant term I_n, which appears because actuator $n-1$ contributes to the rotation of link n. If I_n can be eliminated, by design, such that it does not appear in h_{nn-1}, then decoupling may be possible by choosing $g_n = 0$.

Recalling the expression for the off-diagonal element h_{lm} of Equ. (5.16) given as a function of the sets S_{lm} and S_{lm}^*, it is clear that all rotational effects are contributed by the set S_{lm}. Thus the only way to remove the term I_n from h_{nn-1} is by having the interaction between joints n and $n-1$ contain no rotation effects, that is

$$S_{nn-1} = \varnothing \tag{6.1}$$

This condition states that the set of links that are rotated by both actuators n and $n-1$ simultaneously should be null.

In order for Equ. (6.1) to be satisfied, the actuator driving link n should be relocated in order for the serial drive manipulator of Figure 6-1 to have decoupled motions between actuators n and $n-1$. In other words, actuator n driving link n should be relocated such that actuator $n-1$ induces no rotation to link n.

This process can be repeated so that we arrive to the necessary condition of decoupling for planar serial drive mechanism:

> **Theorem (6.1):** The necessary condition for decoupling the inertia tensor of a planar open loop kinematic chain is

$$S_{lm} = \varnothing, \quad l \neq m. \tag{6.2}$$

In other words, any two different actuators cannot induce rotation to the same link, or more simply, the rotation of each link should be induced by one and only one actuator:

$$S_i = \{i\} \tag{6.3}$$

This makes sense from a physical point of view. If we consider the serial drive planar mechanism of Figure 6-1; when link n accelerates under motion of actuator n at joint n, an induced reaction torque must be borne by actuator $n-1$ at joint $n-1$. This torque is completely dictated by the rotational motion of link *n.*

An important point to note is that the decoupling condition using the set notation, $S_{lm} = \emptyset$ or $S_i = \{i\}$, is a functional relationship. This permits the specification of the general function of a mechanism rather than listing all possible designs that are required to meet the necessary conditions for decoupling.

6.2.2 Sufficient Conditions for a Decoupled and Invariant Inertia Tensor

Consider an open loop kinematic chain consisting of links 1 through n, driven by n independent actuators located at the base frame. The sufficient condition that the manipulator inertia tensor is decoupled and invariant for all arm configurations is given by:

i) $\mathbf{K} = $
$$
\begin{bmatrix}
1 & 0 & 0 & 0 & \cdots & 0 & 0 \\
-1 & 1 & 0 & 0 & \cdots & 0 & 0 \\
0 & -1 & 1 & 0 & \cdots & 0 & 0 \\
0 & 0 & -1 & 1 & \cdots & 0 & 0 \\
& & & \ddots & & & \\
0 & 0 & 0 & 0 & \cdots & -1 & 1
\end{bmatrix}
$$

ii) $\mathbf{r}_{ic}^{i-1} = -\dfrac{1}{m_i} \displaystyle\sum_{j=i+1}^{n} m_j \mathbf{r}_i^{i-1}, \qquad \text{for } i = 2, \dots, n$

where $\mathbf{r}_{ic}^{i-1}$ is the position vector from the origin of coordinate frame $i-1$ (on joint i) to the mass center of link i, $\mathbf{r}_i^{i-1}$ is the position vector from the origin of frame $i-1$ to the origin of frame i, and m_i is the mass of link i.

Proof

The transmission matrix $\mathbf{K}$, as given by (i), is a matrix that corresponds to $S_i = \{i\}$ for all $i = 1, 2, \dots, n$ and thus $S_{lm} = \emptyset$. The angular velocity vector $\mathbf{w}_i$ is given by

$$\mathbf{w}_i = \mathbf{b}_{i-1}\dot{\alpha}_i$$

Note that the partial rate $\mathbf{a}_{il}$ and the joint axis $\mathbf{b}_i$ are all parallel in this case. The linear velocity vector $\mathbf{v}_i$, as given by Equ. (5.12), becomes

$$\mathbf{v}_i = \sum_{j=1}^{i-1} \dot{\alpha}_j \mathbf{b}_{j-1} \times \mathbf{r}_j^{j-1} + \dot{\alpha}_i \mathbf{b}_{i-1} \times \mathbf{r}_{ic}^{i-1}$$

Using (ii), the linear velocity $\mathbf{v}_i$ becomes

$$\mathbf{v}_i = \sum_{j=1}^{i-1} \dot{\alpha}_j \mathbf{b}_{j-1} \times \mathbf{r}_j^{j-1} \frac{1}{m_i} \dot{\alpha}_i \sum_{j=i+1}^{n} m_j \mathbf{b}_{i-1} \times \mathbf{r}_i^{i-1}$$

the kinetic energy is KE,

$$2KE = \sum_{i=1}^{n} m_i \mathbf{v}_i^T \mathbf{v}_i + I_i \mathbf{w}_i^T \mathbf{w}_i$$

$$= \sum_{i=1}^{n} m_i \sum_{j=1}^{i-1} \sum_{k=1}^{i-1} (\mathbf{b}_{j-1} \times \mathbf{r}_j^{j-1})^T (\mathbf{b}_{k-1} \times \mathbf{r}_k^{k-1}) \dot{\alpha}_j \dot{\alpha}_k$$

$$- 2\sum_{i=1}^{n} m_i \sum_{j=1}^{i-1} \dot{\alpha}_j \mathbf{b}_{j-1} \times \mathbf{r}_j^{j-1} \frac{1}{m_i} \dot{\alpha}_i \times \sum_{l=i+1}^{n} m_l \mathbf{b}_{i-1} \times \mathbf{r}_i^{i-1}$$

$$+ \sum_{i=1}^{n} m_i \frac{1}{m_i^2} \dot{\alpha}_i^2 (\sum_{l=i+1}^{n} m_l)^2 \times (\mathbf{b}_{i-1} \times \mathbf{r}_i^{i-1})^T (\mathbf{b}_{i-1} \times \mathbf{r}_i^{i-1})$$

$$+ \sum_{i=1}^{n} \mathbf{b}_{i-1}^T I_i \mathbf{b}_{i-1} \dot{\alpha}_i^2$$

The inner product of two vector products can be expanded as

$$(\mathbf{b}_{j-1} \times \mathbf{r}_j^{j-1})^T (\mathbf{b}_{k-1} \times \mathbf{r}_k^{k-1}) =$$
$$(\mathbf{b}_{j-1}^T \mathbf{b}_{k-1})(\mathbf{r}_j^{j-1}{}^T \mathbf{r}_k^{k-1}) \sim$$
$$(\mathbf{b}_{j-1}^T \mathbf{r}_k^{k-1})(\mathbf{r}_j^{j-1}{}^T \mathbf{b}_{k-1})$$

Since the mechanism is planar, $\mathbf{b}_{j-1}$ is perpendicular to $\mathbf{r}_k^{k-1}$, and

$$(\mathbf{b}_{j-1} \times \mathbf{r}_j^{j-1})^T (\mathbf{b}_{k-1} \times \mathbf{r}_k^{k-1}) = (\mathbf{r}_j^{j-1})^T \mathbf{r}_k^{k-1})$$

$$= |\mathbf{r}_j^{j-1}| \cdot |\mathbf{r}_k^{k-1}| \cos \beta_{jk}$$

$$= L_j \cdot L_k \cos \beta_{jk}$$

$$= C_{jk}$$

where L_j is the length of the vector $\mathbf{r}_j^{j-1}$ and β_{jk} is the angle between link j and link k. The kinetic energy now becomes

$$
2KE = \sum_{i=1}^{n} m_i \sum_{j=1}^{i-1} \sum_{k=1}^{i-1} C_{jk} \dot{\alpha}_j \dot{\alpha}_k
$$
$$
- 2 \sum_{i=1}^{n} \sum_{j=1}^{i-1} (\sum_{l=i+1}^{n} m_l) C_{ji} \dot{\alpha}_j \dot{\alpha}_i
$$
$$
+ \sum_{i=1}^{n} \left[\frac{1}{m_i} (\sum_{l=i+1}^{n} m_l)^2 C_{ii} + \mathbf{b}_{i-1} \mathbf{I}_i \mathbf{b}_{i-1} \right] \dot{\alpha}_i^2
$$

and by switching the order of summations, we obtain

$$
2KE = \sum_{j=k+1}^{n} (\sum_{i=\max[j,k]+1}^{n} m_i) C_{jk} \dot{\alpha}_j \dot{\alpha}_k + \sum_{j=1}^{n} (\sum_{i=j+1}^{n} m_i) C_{jj} \dot{\alpha}_j^2
$$
$$
- 2 \sum_{j=1}^{n} \sum_{i=j+1}^{n} (\sum_{l=i+1}^{n} m_l) C_{ji} \dot{\alpha}_j \dot{\alpha}_i
$$
$$
+ \sum_{i=1}^{n} \left[\frac{1}{m_i} (\sum_{l=i+1}^{n} m_l]^2 C_{ii} + \mathbf{b}_{i-1}^T \mathbf{I}_i \mathbf{b}_{i-1} \right] \dot{\alpha}_i^2
$$

$$
= \left[\sum_{i=1}^{n} (\sum_{l=i+1}^{n} m_l)(\frac{1}{m_i})(\sum_{l=i+1}^{n} m_l + 1) C_{ii} + \mathbf{b}_{i-1}^T \mathbf{I}_i \mathbf{b}_{i-1} \right] \dot{\alpha}_i^2
$$
$$
+ 2 \sum_{j=1}^{n} \sum_{k=j+1}^{n} (\sum_{l=k+1}^{n} m_l) C_{jk} \dot{\alpha}_j \dot{\alpha}_k
$$
$$
- 2 \sum_{j=1}^{n} \sum_{k=j+1}^{n} (\sum_{l=k+1}^{n} m_l) C_{jk} \dot{\alpha}_j \dot{\alpha}_k
$$

or

$$=[\sum_{i=1}^{n}(\sum_{l=i+1}^{n}m_l)\frac{1}{m_i}(\sum_{l=i+1}^{n}m_l)\cdot a_i^2+I_i]\ddot{\alpha}_i^2$$

thus the elements of the inertia tensor are given by

$$h_{jk}=0;\ \forall\ j\neq k$$

$$h_{ii}=\sum_{i=1}^{n}(\sum_{l=i+1}^{n}m_l)\frac{1}{m_i}(\sum_{l=i}^{n}m_l)\cdot a_i^2+I_i,\quad i\in\{2,3,\ldots,n\}$$

Thus the inertia tensor is completely decoupled and invariant for all arm configurations. **Q.E.D.**

6.3 Parallelogram Mechanisms

This section explores the differences in dynamic behavior that exist between two 2 d.o.f parallel drive mechanisms that use either a five-bar-link mechanism or a chain. The two mechanisms examined first use parallel drive as shown in Figures 6-2 and 6-3 The set notation was applied to these two mechanisms in examples 3 and 4 of Chapter 5. It was found that the elements of the inertia tensor of the mechanism driven through a chain, Figure 6-2, depend on the following mass properties

$$h_{11}=h_{11}(m_1,m_4,I_1)$$

$$h_{22}=h_{22}(m_4,I_4)$$

$$h_{12}=h_{12}(m_4)$$

and they are explicitly given by

$$h_{11} = I_1 + m_1 g_1^2 + m_4 l_1^2$$

$$h_{22} = I_4 + m_4 g_4^2$$

$$h_{12} = m_4 l_1^2 g_4 \cos(\alpha_2 - \alpha_1)$$

On the other hand, the set notation was also applied to the mechanism which uses a five-bar-link mechanism in Chapter 5, to determine the mass properties that are involved in the elements of the inertia tensor. These implicit dependences are reproduced below

$$h_{11}^* = h_{11}(m_1, m_3, m_4, I_1, I_3)$$

$$h_{22}^* = h_{22}(m_4, m_2, m_3, I_2, I_4)$$

$$h_{12}^* = h_{12}(m_3, m_4)$$

or explicitly,

$$h_{11}^* = I_1 + m_1 g_1^2 + I_3 + m_3 g_3^2 + m_4 l_1^2$$

$$h_{22}^* = I_4 + m_4 g_4^2 + I_2 + m_2 g_2^2 + m_3 l_2^2$$

$$h_{12}^* = (m_3 l_2 g_3 - m_4 l_1 g_4) \cos(\alpha_2 - \alpha_1)$$

Let us define a vector $\mathbf{p}^T = (m_2 \ m_3)$ whose components are the masses of links 2 and 3, and let $I_i = m_i R_i^2$ for $i = 2,3$ where R_i is the radius of gyration of link i, then[3]

---

[3]Proper angles are used before taking the limit.

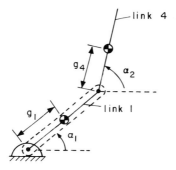

Figure 6-2: Parallel drive mechanism using a chain

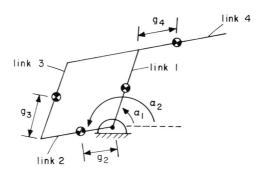

Figure 6-3: Parallel drive mechanism using a five-bar mechanism

$$\lim_{\mathbf{p}\to 0} h_{11}^{*} = h_{11}$$

$$\lim_{\mathbf{p}\to 0} h_{22}^{*} = h_{22}$$

$$\lim_{\mathbf{p}\to 0} h_{12}^{*} = h_{12}$$

The two inertia tensors are similar . Thus, the extra links merely transmit the motion from the actuator to the forearm, link 4, as expected.

As a counter example, consider the planar mechanism of Figure 3-4, where the actuators are at the base. There is no mass distribution that can decouple the inertia tensor of this mechanism since the angular velocity vector of link 3 is affected by both actuators. This example shows that simply placing the actuators at the base is not sufficient to achieve decoupling of the manipulator inertia tensor.

6.4 Decoupling of Serial Spatial Mechanisms with Actuator Relocation

The class of manipulators of interest are still the open kinematic chain type where the actuators can be located anywhere along the linkage. The goal is to utilize the fundamental result on decoupling, presented in Chapter 4, namely that the reaction torques due to centroidal rotation must be eliminated. Thus to meet this binding necessary condition, it suffices to eliminate all of the rotational effects in each off-diagonal element, h_{lm}, of the manipulator inertia tensor. The links i that contribute moments of inertia to the element h_{lm} belong to the set S_{lm} as derived previously. The centroidal rotation effects, as given in Equ. (5.16), are reproduced below

$$(6.4)$$

$$h_{lm}^{\text{rot}} = \sum_{i \in S_{lm}} \mathbf{a}_l^T \mathbf{I}_i \mathbf{a}_{im}$$

To eliminate this term completely, we relocate the actuators and choose the mass properties of certain links so that : (i) the set S_{lm} is the null set or; (ii) the entire sum is zero.

Design Guideline:

1. $S_{lm} = \varnothing$, or

2. $\mathbf{a}_{il}^t \mathbf{I}_i \mathbf{a}_{im} = 0;$ $\quad \forall \, i \in S_{lm}$, and for all arm configurations.

Proof

The first design guideline states that the set of links that are rotated by both actuators l and m must be the null set. Since we are interested in the off-diagonal element h_{lm}, $l \neq m$, one way to achieve this is by having the rotation of each link be induced by one and only one actuator, $S_i = \{i\}$. Using the definition given by Equ. (5.17), it follows that $S_{lm} = \varnothing$, and no centroidal rotations are involved in h_{lm}, since h_{lm}^{rot} is zero.

One way to satisfy the second design rule, is for each link i that is rotated by both actuators l and m, we choose either partial rate of orientation $\mathbf{a}_{il}$ or $\mathbf{a}_{im}$ to be a principal direction for the inertia tensor $\mathbf{I}_i$ of link i for all arm configurations; and (ii) the partial rates of orientation $\mathbf{a}_{il}$ and $\mathbf{a}_{im}$, of actuators l and m that both induce rotation to link i, to be orthogonal for all arm configurations, namely

i) For all $i \in S_{lm}$, $\mathbf{a}_{il}$ or $\mathbf{a}_{im}$ be a principal direction[4] for link i, and

ii) $\mathbf{a}_{il}^T \mathbf{a}_{im} = 0$ for $l \in S_i$ and $m \in S_i$ for all arm configurations.

In other words, using the definition of the set S_i, this is equivalent to saying that all partial rate vectors, $\mathbf{a}_{il}$, that make up the angular velocity vector $\mathbf{w}_i$ of any link i must be mutually orthogonal. This design rule permits the elimination of all moments of inertia from the off-diagonal element h_{lm} in order to satisfy the necessary condition for decoupling serial spatial mechanisms.

Without loss of generatlity, we can assume that $\mathbf{a}_{im}$ is in a principal direction of $\mathbf{I}_i$. Let us assume a coordinate transformation so that the direction of $\mathbf{a}_{im}$ coincides with the z-axis of the new coordinate frame, then the inertia tensor can be expressed as

---

[4]In general, there is no rotational effect if $\mathbf{a}_{il}$ and $\mathbf{a}_{im}$ are $\mathbf{I}_i$ orthogonal for $l \neq m$

$$\mathbf{I}_i = \begin{bmatrix} I_{xx} & I_{xy} & 0 \\ I_{xy} & I_{yy} & 0 \\ 0 & 0 & I_{zz} \end{bmatrix}$$

Now we consider the quadratic form $\rho = \mathbf{a}_{il}^T \mathbf{I}_i \mathbf{a}_{im}$, where $\mathbf{a}_{im}$ is a principal axis for $\mathbf{I}_i$ and the partial rates are given by,

$$\mathbf{a}_{il}^T = (a\,b\,c)$$

$$\mathbf{a}_{im}^T = (\alpha\,\beta\,\gamma) = (0\,0\,\gamma)$$

then

$$\mathbf{a}_{il}^T \mathbf{I}_i \mathbf{a}_{im} = c\,\gamma I_{zz}$$

since $\mathbf{a}_{il}^T \mathbf{a}_{im} = 0$, then $\rho = 0$ and no moments of inertia should appear.

6.5 Conclusion

This Chapter provided some design conditions for decoupling using the results of the theorem and corollaries of Chapter 4 and the set notation introduced in Chapter 5. The reaction torque condition was expressed as a design rule for decoupling spatial serial mechanisms in terms of the partial rates of angular velocity and the sets S_i and S_{lm}. This was utilized in deriving the necessary and sufficient conditions for a decoupled and invariant inertia tensor of an n degree of freedom planar mechanisms, and providing a design guideline for spatial mechanisms.

Part III

DESIGN APPLICATION - M.I.T. DIRECT-DRIVE MANIPULATORS

Chapter 7

Mechanisms

7.1 Introduction

In this chapter, we present the design of the M.I.T. 3 d.o.f. direct-drive manipulators. Specifically, we introduce the arm mechanisms and then derive the mass redistribution conditions for which their inertia tensors are decoupled and invariant. The planar dynamics, however, are analyzed first.

7.2 Arm Mechanism

The three-degree-of-freedom direct-drive arms to be analyzed are models II and III and are shown in Photos 1-3 and 1-4. These models show a parallel drive mechanism consisting of a five-bar-link and a six-bar-link mechanism that forms a parallelogram. Reference [Asada and Youcef-Toumi 84] describes the first direct-drive arm prototype that uses a parallel drive mechanism. Figure 7-1 shows the schematic design for model II. The schematic of the M.I.T. D.-D. Arm model III of Photo 1-4 is shown in Figures 7-2 and 7-3. As seen in Figure 7-2, when joint 2 rotates link 2, the arm tip performs a reach motion. A lift motion is achieved by having joint 3 rotate link 3. A vertical planar motion of the arm tip can be accomplished by actuating joints 2 and 3 together. Finally, the base joint, joint 1, rotates the whole mechanism about the vertical axis allowing a three dimensional motion of the arm tip. This arm [Asada, Youcef-Toumi and Ramirez 84, Ramirez 84] uses the key feature of the parallel drive, as presented in [Asada and Youcef-Toumi 83a, Asada and Youcef-Toumi 84], which allows the decoupling of the inertia tensor by mass distribution.

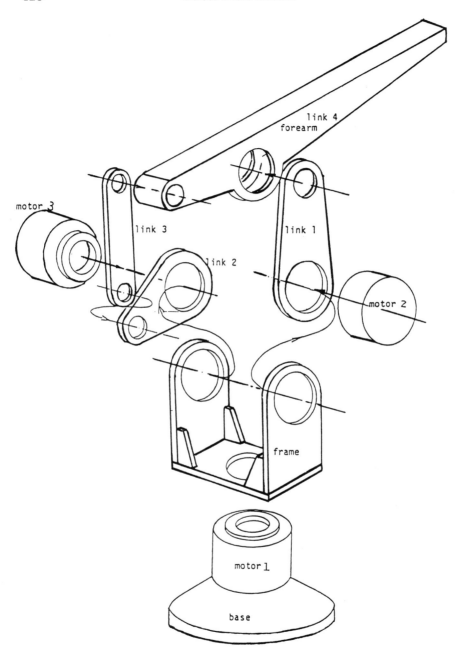

Figure 7-1:   Schematic of the M.I.T. 3 d.o.f direct-drive manipulator:
model II

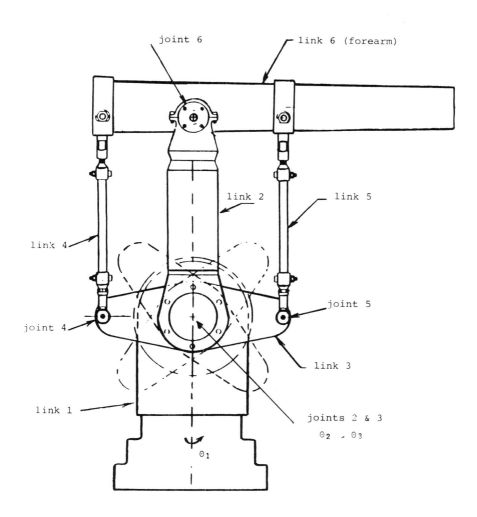

Figure 7-2: Schematic of the M.I.T. 3 d.o.f direct-drive manipulator
model III: side view

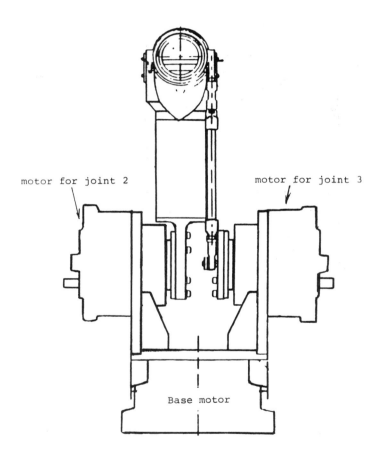

motor for joint 2

motor for joint 3

Base motor

Figure 7-3: Schematic of the M.I.T. 3 d.o.f direct-drive manipulator
model III: front view

7.3 Decoupling of Planar Dynamics

Let us immobilize the base joint of the manipulator shown in Figure 7-1 so that joints 2 and 3 are the only active joints. This mechanism is equivalent to the five-bar-link mechanism described in Chapter 6, where we showed that the five-bar-link mechanism is equivalent to the parallel drive which uses a chain mechanism. Therefore, the necessary condition for decoupling should apply. In what follows we derive the conditions that must be satisfied so that the manipulator inertia tensor becomes decoupled and/or invariant.

7.3.1 Five-Bar-Link Mechanisms

In this section, we design an arm structure and its mass properties so that the arm has an invariant and decoupled inertia tensor which causes no nonlinear velocity torques. Specifically, we derive the conditions that make the off-diagonal element H_{ij} zero in Equ. (4.1). The arm dynamics are then extremely simple having only the first and the last terms of Equ. (4.2).

Figure 7-4 shows a special case of the parallel drive five-bar-link mechanism, in which the distance between the two motors is zero and the two pairs of opposite links are parallel. First we immobilize motor 2 and compute the moment of inertia about the axis of motor 1, which corresponds to H_{11} in Equ. (4.1). Inertia H_{11} is the resultant inertia of link 1, link 3 and link 4 about joint 3. The inertia of link 1 about the first motor axis is given by $I_1 + m_1 g_1^2$, which is invariant. Since link 3 rotates about joint 3 when link 1 rotates, the resultant inertia of link 3 involved in H_{11} is $I_3 + m_3 g_3^2$, which is also invariant for all arm configurations. Similarly, the mass center of link 4 moves along the circle of radius l_1, keeping constant attitude. Therefore, the resultant inertia of link 4, which is also invariant and configuration independent, appears as $m_4 l_1^2$ in H_{11}. Thus the overall inertia about the second motor axis, when motor 1 is immobilized, is also invariant by the same reason. The inertia tensor of the arm mecnanism is them given by

$$\mathbf{H} = \begin{bmatrix} H_{11} & H_{12} \\ H_{12} & H_{22} \end{bmatrix}$$

where

$$H_{11} = I_1 + m_1 g_1^2 + I_3 + m_3 g_3^2 + m_4 l_1^2$$

$$H_{22} = I_4 + m_4 g_4^2 + I_2 + m_2 g_2^2 + m_3 l_2^2 \qquad (7.1)$$

$$H_{12} = \left(m_3 l_2 g_3 - m_4 l_1 g_4 \right) \cos \left(\theta_2 - \theta_1 \right)$$

Off-diagonal element H_{12} governs the interaction between the two motors. When the angle between link 1 and link 4 is a right angle, H_{12} becomes zero, then the arm has no interaction. As the angle diverges from this angle, the interaction problem becomes prominent.

Instead of depending on the arm configuration, it is possible to eliminate the interaction completely for all configurations by modifying the dimensions and mass properties of the arm. In the above expression for H_{12}, we can modify the mass ratio of link 3 and link 4, m_4/m_3, the ratio of mass center distances of the two links, g_4/g_3, and/or the length l_2/l_1 so that the coefficient $(m_3 l_2 g_3 - m_4 l_1 g_4)$ becomes zero. The condition for which the arm has no interactions for all configurations is given by

$$\frac{m_4}{m_3} \cdot \frac{g_4}{g_3} = \frac{l_2}{l_1} \qquad (7.2)$$

The resultant inertia tensor which satisfies the above condition is reduced to

$$\mathbf{H} = \begin{bmatrix} I_1 + I_3 + m_1 g_1^2 \\ + m_4 l_1^2 \left(1 + \dfrac{g_3 g_4}{l_1 l_2}\right) & 0 \\ & I_2 + I_4 + m_3 l_2^2 \\ 0 & + m_4 g_4^2 \left(1 + \dfrac{l_1 l_2}{g_3 g_4}\right) \end{bmatrix} \qquad (7.3)$$

The inertia tensor is invariant and completely decoupled. Therefore, there are no variations in motor load, no interactions, and no Coriolis and centrifugal forces.

In the case of the serial drive mechanism in Figure 7-5, the inertia tensor is given by

$$H_{11} = I_1 + m_1 g_1^2 + I_4 + m_4 \left(l_1^2 + g_4^2 - 2 l_1 g_4 \cos \theta_2 \right)$$

$$H_{22} = I_4 + m_4 g_4^2 \tag{7.4}$$

$$H_{12} = I_4 + m_4 g_4^2 - m_4 l_1 g_4 \cos \theta_2$$

Since inertia H_{11} depends on θ_2, it is configuration dependent. The off-diagonal element H_{12} is also configuration dependent and cannot be reduced to zero for all θ_2 by changing the mass properties. It is impossible to eliminate the interactions for the serial drive mechanism. Thus, the parallel drive mechanism mentioned above has significant advantages over the serial drive. The inertia tensor can be made invariant if the forearm is balanced so that $g_4 = 0$.

7.3.2 Six-Bar-Link Mechanisms

Using the notation of Figure 7-6, the sets representing rotations are

$$S_2 = S_4 = S_5 = \{2\} \quad \text{and} \quad S_3 = S_6 = \{3\}$$

The set $S_{23} = \varnothing$, thus no link moments of inertia are involved in the off diagonal element h_{23}. The necessary condition for decoupling a planar mechanism, is then satisfied. The elements of the inertia tensor for the two-degree-of-freedom arm are given by

$$\mathbf{G}_{23} = \begin{bmatrix} h_{22} & h_{23} \\ h_{23} & h_{33} \end{bmatrix}$$

where its elements are given by

$$h_{22} = I_2 + m_2 g_2^2 + I_4 + m_4 g_4^2 + I_5 + m_5 g_5^2 + m_6 l_2^2 \tag{7.5}$$

$$h_{33} = I_3 + m_3 g_3^2 + m_4 l_3^2 + m_5 l_3^2 + I_6 + m_6 g_6^2$$

$$h_{23} = (m_5 l_3 g_5 + m_6 l_2 g_6 - m_4 l_3 g_4) \cos(\alpha_3 - \alpha_2)$$

The parameters m_i, I_i represent the mass and mass moment of inertia about

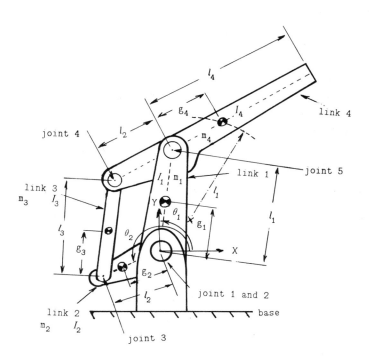

Figure 7-4: 2 d.o.f. Mechanism using a Five-bar-link Parallel drive:
Dimensions and Mass Properties

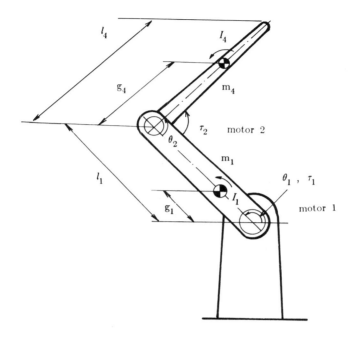

Figure 7-5: 2 d.o.f. Mechanism using a serial drive:
Dimensions and Mass Properties

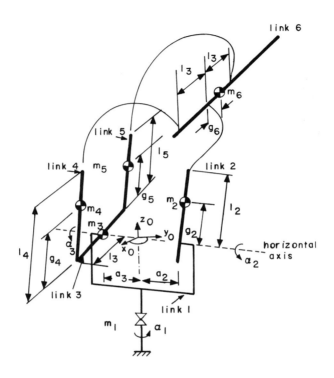

Figure 7-6: Skeleton of the 3 d.o.f M.I.T. direct-drive Arm model III

an axis going through the mass center of link i and parallel to the horizontal axis shown in Figure 7-6; l_i and g_i are its link length and the distance to the center of mass as defined in Figure 7-6.

It is apparent from Equ. (7.5) that the diagonal elements of the inertia tensor, h_{22} and h_{33}, do not involve actuator displacements α_2 and α_3, hence, they are invariant for all arm configurations. The coupling term h_{23} is caused mainly by the motion of the mass centers of links 4, 5 and 6. This term can be made equal to zero for all arm configurations when the following is satisfied,

$$m_5 l_3 g_5 + m_6 l_2 g_6 - m_4 l_3 g_4 = 0$$

    or

$$\frac{m_5}{m_4} \cdot \frac{g_5}{g_4} + \frac{m_6}{m_4} \cdot \frac{l_2}{l_3} \cdot \frac{g_6}{g_4} = 1$$

(7.6)

The above equation suggests the redistribution of the mass of links 4, 5 and 6 in order to decouple the arm inertia tensor. This can be achieved by changing the mass center locations, g_4, g_5 and g_6, or by changing the mass proportions of the three links; $m_4:m_5:m_6$. One possible solution to Equ. (7.6) occurs when links 4 and 5 are identical so that $m_5\, l_3\, g_5$ and $m_4\, l_3\, g_4$ cancel each other. The decoupling condition then reduces to

$$g_6 = 0$$

(7.7)

Under this condition the mass center of the forearm coincides with joint 6. This is equivalent to static mass balancing of the forearm. For the six-bar-link mechanism, this is a condition for decoupling the inertia tensor.

In the case of the five-bar-link mechanism of Figure 7-1, the sets representing rotation and the elements of its inertia tensor were already derived in Chapter 6. The condition for achieving a decoupled and invariant inertia tensor was previously found to be

$$\frac{m_4}{m_3} \cdot \frac{g_4}{g_3} \cdot \frac{l_1}{l_2} = 1$$

7.4 Decoupling of Spatial Dynamics

The arm inertia tensor depends not only on the mass properties of each arm link but also on the kinematic structure of the arm linkage. Thus, the kinematic structure is critical for achieving a decoupled inertia tensor. For the manipulator shown in Figures 7-2 and 7-3, the parallelogram mechanism was used for decoupling the upper two joints, 2 and 3. It is also possible to decouple the arm inertia tensor in the 3 d.o.f case for all arm configurations. In this case, we use the design guideline of Chapter 6 for spatial mechanisms, that is to eliminate the quadratic forms $a_{il} I_i a_{im}$. The angular velocities of the links are

$$w_1 = e_1 \dot{\alpha}_1$$

$$w_2 = w_4 = w_5 = e_1 \dot{\alpha}_1 + e_2 \dot{\alpha}_2$$

$$w_3 = w_6 = e_1 \dot{\alpha}_1 + e_2 \dot{\alpha}_3$$

where Kane's partial rates e_1 and e_2 are

$$e^T_1 = (0 \quad 0 \quad 1) \text{ and } e^T_2 = (\cos \alpha_1 \quad \sin \alpha_1 \quad 0)$$

Note that the vectors e_1 and e_2 are orthogonal. Also, the principal direction of each link of the upper arm is in the direction of e_2. Therefore, the necessary condition is satisfied since we eliminated all the moments of inertia from the off-diagonal elements of the manipulator inertia tensor.

In Figure 7-6, note that the axis of joint 1 is perpendicular to the axes of joints 2 and 3. Therefore, torques exerted by joints 2 and 3 would not act upon joint 1 for a balanced arm design. The elements of the inertia tensor of the 3 d.o.f manipulator are given by

$$h_{11} = I^1{}_{zz} + m_2(a^2{}_2 + g^2{}_2 cos^2\alpha_2) + I^2{}_{zz} sin^2\alpha_2 + I^2{}_{yy} cos^2\alpha_2 + m_3 a^2{}_3$$

$$+ I^3{}_{yy} sin^2\alpha_3 + I^3{}_{zz} cos^2\alpha_3$$

$$+ m_4(a^2{}_3 + I^2{}_3 cos^2\alpha_3 + g^2{}_4 cos^2\alpha_2) + I^4{}_{yy} cos^2\alpha_2 + I^4{}_{zz} sin^2\alpha_2$$

$$+ m_5(a^2{}_3 + I^2{}_3 cos^2\alpha_3 + g^2{}_5 cos^2\alpha_2) + I^5{}_{yy} cos^2\alpha_2 + I^5{}_{zz} sin^2\alpha_2$$

$$+ m_6(a^2{}_2 + I^2{}_2 cos^2\alpha_2 + g^2{}_6 cos^2\alpha_3) + I^6{}_{yy} sin^2\alpha_3 + I^6{}_{zz} cos^2\alpha_3$$

$$+ 2I^6{}_{yz} sin\alpha_3 cos\alpha_3$$

$$- 2m_4 l_3 g_4 cos\alpha_2 cos\alpha_3 + 2m_5 l_3 g_5 cos\alpha_2 cos\alpha_3$$

$$+ 2m_6 l_2 g_6 cos\alpha_2 cos\alpha_3$$

$$h_{22} = I^2{}_{xx} + m_2 g^2{}_2 + I^4{}_{xx} + m_4 g^2{}_4 + I^5{}_{xx} + m_5 g^2{}_5$$

$$+ m_6 l^2{}_2$$

$$h_{33} = I^3{}_{xx} + m_4 l^2{}_3 + m_5 l^2{}_3 + m_6 g^2{}_6 + I^6{}_{xx}$$

$$h_{12} = (m_4 g_4 a_3 + m_5 g_5 a_3 - m_6 l_2 a_2 - m_2 g_2 a_2) sin(\alpha_2)$$

$$h_{13} = (-m_4 l_3 a_3 + m_5 l_3 a_3 - m_6 g_6 a_2) sin(\alpha_3)$$

$$h_{23} = (-m_4 l_3 g_4 + m_5 l_3 g_5 + m_6 l_2 g_6) cos(\alpha_3 - \alpha_2)$$

where $I^i{}_{xy}$ is the x–y moment of inertia of link i. As expected, no moments of inertia should be involved in the off-diagonal elements. The base joint inertia, h_{11}, is a function of the actuator displacements α_2 and α_3 as expected. The base motor would experience a high inertia when the upper arm is extended and a low inertia when it is retracted. The terms h_{22} and h_{33} are constants since their corresponding inertia are not affected by any of the joint motions. This results from the use of the parallel drive, as discussed previously. The off-diagonal elements are functions of configuration, however , they can be eliminated. The corresponding mass redistribution conditions that make the inertia tensor decoupled for all arm configurations are

$$m_4 g_4 a_3 + m_5 g_5 a_3 - m_6 l_2 a_2 - m_2 g_2 a_2 = 0 \tag{7.8}$$

$$-m_4 l_3 a_3 + m_5 l_3 a_3 - m_6 g_6 a_2 = 0$$

$$-m_4 l_3 g_4 + m_5 l_3 g_5 + m_6 l_2 g_6 = 0$$

The acceleration of any link produces both translational inertial forces and rotational inertial forces at its mass center. The former depends only on the mass of the link and the latter on the moment of inertia. These conditions (7.8) are designed only to eliminate interactive torques resulting from the linear motion of the mass center of each link. The effects due to rotation are not present because we used design guideline (ii). The motion of each link, in the upper arm, is restricted to a vertical plane; as a result any acceleration torque resulting from rotation is orthogonal to the base joint axis and does not cause interactions. Another result is that static balancing of all the links by making all g_i equal to zero does not alter the space dependency of the elements h_{12} and h_{13}.

In the design of this particular arm links 4 and 5 were made the same, thus the decoupling conditions become

$$2m_4 g_4 a_3 - (m_6 l_2 + m_2 g_2) \, a_2 = 0$$

$$g_6 = 0$$

If we balance link 2 such that g_2 is zero, then the conditions reduce to

$$\frac{m_6}{m_4} \cdot \frac{l_2}{g_4} \cdot \frac{a_2}{a_3} = 2 \tag{7.9}$$

$$g_2 = 0, \quad g_6 = 0$$

For the planar dynamics of the upper arm, the governing equation of each actuator does not include interactive acceleration torques, or nonlinear velocity torques since the inertia tensor G_{23} is decoupled and invariant . In the 3-d.o.f case, the inertia tensor is completely decoupled. However, the inertia of the base motor is a function of actuator displacements α_2 and α_3, hence some velocity torques are present. The resultant simplified dynamic equations are

$$h_{11}(\alpha_2,\alpha_3)\ddot{\alpha}_1 + \frac{\partial h_{11}}{\partial \alpha_2}\dot{\alpha}_1\dot{\alpha}_2 + \frac{\partial h_{11}}{\partial \alpha_3}\dot{\alpha}_1\dot{\alpha}_3 = \tau_1$$

$$h_{22}\ddot{\alpha}_2 - \frac{1}{2}\frac{\partial h_{11}}{\partial \alpha_2}\dot{\alpha}_1^2 + \tau_{g2} = \tau_2$$

$$h_{33}\ddot{\alpha}_3 - \frac{1}{2}\frac{\partial h_{11}}{\partial \alpha_3}\dot{\alpha}_1^2 + \tau_{g3} = \tau_3$$

7.5 Conclusion

This chapter has addressed the design of the M.I.T direct-drive manipulators with decoupled inertia tensors by mass redistribution. The design guidelines of Chapter 6 were used to make the decoupling possible.

Chapter 8

Control Systems

8.1 Introduction

This chapter explores the overall control performance of the prototype direct-drive manipulator. Analytical and experimental evaluations are conducted for speed, acceleration, servo stiffnesses, actuator interactions and the torque control capability of the direct-drive robot.

8.2 Drive System Hardware

8.2.1 Actuators

The performance of the direct-drive arm is highly dependent on the performance of the motors at the active joints. An investigation has been conducted on actuators for the direct-drive arm [Asada,1981] . The motors used in this particular design are brushless DC motors which utilize a new type of rare-earth-cobalt permanent magnet. These magnets have an extremely large maximum magnetic energy product, $BH_{max} = 26 MGOe$. Since the magnets are not demagnetized in a normal operating range, a large current can be applied to the armature of the motor resulting in a large torque which can accelerate (decelerate) the arm joints very rapidly.

Figure 8-1 shows the brushless motor housed in a case. The rotor, which contains the magnets, is directly coupled to the joint axis. A magnetic brake and a tachometer generator with hallow shafts are also mounted in the case which is aligned with the joint axis. The windings are part of the stator (not the rotor) consequently the motor has a better heat dissipation than a DC torque motor [Asada,1981] because the heat generated is efficiently dissipated through the motor housing. Also, brushless motors can use high currents without causing sparks or mechanical wear .

Table 8-1 lists the characteristics of the motors used in the M.I.T direct-

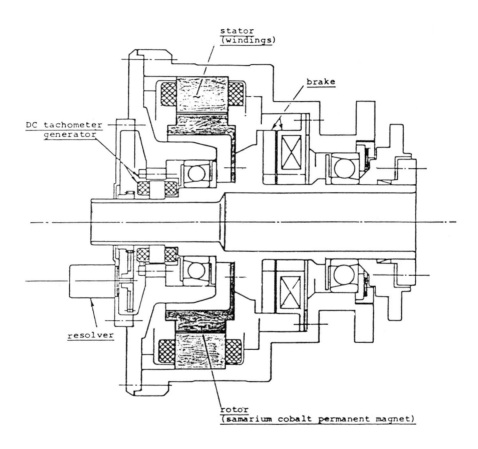

Figure 8-1: Brushless DC Motor

drive arms models II and III. These characteristics are an order of magnitude better than the DC torquers in reference [Asada,1981] .

motor diameter (cm)	25	35
peak torque (Nm)	230	660
rotor inertia (Kg-m^2)	0.0256	0.181
number of poles	18	30
maximum instantaneous current (Amps)	15	50
maximum continuous current (Amps)	10	15
mass (Kg)	16.5	20.39
motor constant Nm/$\sqrt{W}$	2.5	7.5

Table 8-1: Rare-Earth Brushless DC Motor Specifications

8.2.2 Sensors

A set of position and velocity sensors are mounted on each motor. A resolver generates two AC signals that are processed by a resolver to digital (R-to-D) converter to output a 16-bit signal representing position. This signal is used for commutating current in the three phases of the motor and also for position feedback. The three phase currents are measured by resistors in series with each of the phase windings. The torque exerted by the motor can be estimated from these currents. The tachometer shown in Figure 8-1 provides a measure of velocity.

8.2.3 Power Amplifier

The driving amplifiers used are Pulse Width Modulated (PWM),"H" bridge type with switching at a rate of 2 KHz . The motor voltage is switched at a high voltage, 340 volts, to provide high currents. The basic block diagram of the drive amplifier was shown in Figure 2-3. The R/D converter feeds a 10-bit signal representing the actual rotor position to the ROM's which output the sine scaling functions for each phase. These functions are in turn multiplied by the current request through the multiplying Digital/Analog converters to output the request current in each phase. Finally, the pulse width modulator and the output power stage work to bring the error between the request currents and the actual phase currents to zero.

8.3 Position Control Evaluation

8.3.1 Actuator Modeling and Identification

The direct coupling of the motor rotors to the load requires motors with high torque output. To achieve such high torque, one must increase the number of turns in the motor windings. However, the inductance is proportional to the square of the number of turns, hence the delay in motor response increases and the control task becomes difficult. The power amplifiers were designed with current feedback to minimize the inductance effects, improve control of the torque output, and avoid the effects of line voltage fluctuations.

The first goal is to evaluate the motor-amplifier combination in the direct-drive application. Figure 8-2 shows a block diagram of the PWM amplifier and its motor [Electro-Craft, Persson 1975 and 1976]. The differential command input ΔV is converted to a desired current by the gain k_1 . The PWM, modelled as a gain k_a, outputs a mean voltage V_m, the back EMF must be subtracted from this to yield an effective voltage V_{eff} across the motor windings. The phase resistance r and inductance l form the winding impedance, $r+ls$, where s stands for the Laplace operator. The motor torque τ_m which is the product of the torque constant and the motor current i accelerates the motor inertia J_m. K_i is a current feedback constant. In this model, lower order effects due to mutual inductance windings, overlapping conduction angles and unequal rise and fall time of currents are neglected. The transfer function relating the input to the amplifier $\Delta V(s)$ and the angular velocity $\omega(s)$ is

$$\frac{\omega(s)}{\Delta V(s)} = \frac{\dfrac{k_1 k_a k_t}{J_m l}}{s^2 + \dfrac{r+k_i k_a}{l}s + \dfrac{k_t k_b}{J_m l}} \qquad (8.1)$$

where k_b is the back EMF constant. The two poles are functions of the current gain k_i.

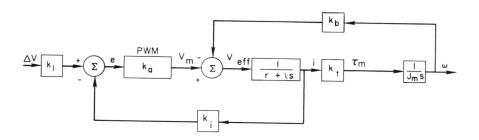

Figure 8-2: Block Diagram of PWM Drive Amplifier and motor

An experimental verification of the fundamental mathematical model presented is necessary to validate the analysis and amplifier design. A frequency response test was conducted on the amplifier-motor combination using a Spectrum Analyzer (H-P 5423A). The input/output variables considered were the differential input voltage to the amplifier and the motor velocity respectively. It was clear that the system has two real poles, thus it is second order. The slowest pole was identified as a load dependent parameter by repeating the frequency response test with different load conditions . The fastest pole was only affected slightly. A typical open loop frequency response is shown in Figure 8-3. Under zero load conditions, the amplifier-motor system had a DC gain of 35 db and two poles at 3.65 Hz and 68 Hz . This verifies the mathematical model for the amplifier-motor combination.

8.3.2 Position Control System Design

Since the prototype arm's inertia is decoupled and nearly invariant, control problems reduce to single-axis control problems with fixed inertial loads. In this section, we present the design for the single-axis controller (refer to Appendix C) and evaluate its control performance.

Since we used large windings to exert a large torque, the motor has a significantly large inductance, which results in a lower frequency at the second pole. Unlike the conventional servomotor where inductance is negligibly small, the high torque brushless motor has a higher delay due to this large inductance, which leads to poor stability when the position loop is closed.

Figure 8-4 shows the step response when motor 1 drives the five-bar-link mechanism and position and velocity are fed back. The response ,with a delay time of 82 ms ,is fairly slow and poor in damping. When acceleration feedback was included, the response was significantly improved as shown in the same Figure. In this case the delay time is only 27 ms. Thus, the feedback of acceleration or equivalent higher order derivatives is necessary for proper control. Instead of using an accelerometer, motor currents monitored at the windings and the quasi-derivative of a tachometer signal were used to get the acceleration signal. The tachometer used was a pancake-type D.C. tachometer which has low ripple, less than 1 % of average reading, and a high sensitivity of 2.2 Volt/rad/s. The servo stiffnesses [Asada, Kanade and Takeyama 83] of the base motor and motors for the five-bar-link mechanism were measured by applying known loads and measuring the resultant displacements at the arm tip. The stiffnesses achieved were 300 N/mm for the base joint and 416 N/mm for

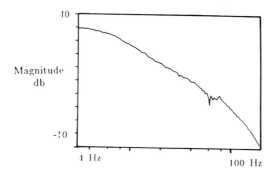

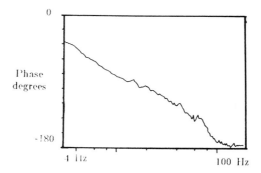

Figure 8-3: Open loop frequency response of Amplifier-motor system

Direct-Drive Robots

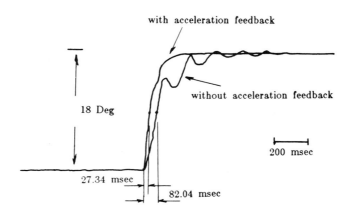

Figure 8-4: Position response of motor 1 driving the link mechanism

the other two joints. These servo stiffnesses, which depend on the loop gains and link lengths, are 10 to 14 times larger than those of the direct-drive arm in reference [Asada, Kanade and Takeyama 83]. The bandwidths (at 3 decibels cutoff when the position loop is closed) were measured to be 3.17 Hz for the base and and 3.1 Hz the other motors respectively. These bandwidths, while measured for large amplitudes of about 90 degrees span, are comparable to the bandwidths of hydraulically driven robots.

Figure 8-5 shows a velocity profile when the base joint traveled a long distance from -90 deg to +90 deg. The arm was extended 0.75 m from the base center with no payload. The speed indicated is that of the end point. The end point was accelerated very rapidly and it reached the top speed. The maximum acceleration determined by the slope is $52\, m/s^2$, which is more than 5 G. Since most industrial robots currently used have at most 0.1 to 0.5 G, the developed direct-drive arm is an order-of-magnitude faster in acceleration. The top speed measured was 9.36 *m/sec*, which is also much faster than that of existing robots.

8.4 Dynamic Decoupling Evaluation

The following experiment was conducted to evaluate the coupling between the motors. One motor, tracking a sinusoidal position command, drives the link mechanism while the other motors are not servoing. The peak-to-peak ratios between the actual position signals of the driving motor and the other motors is a measure of the coupling for the arm dynamics. These ratios (in Db's) are listed in Table 8-2. The test was performed for two frequencies of the driving joint, 0.5 and 5 Hz, and for four different arm configurations. The experimental data shows a coupling of less than -30 Db for all arm configurations. Thus the arm is well decoupled.

8.5 Torque Control

8.5.1 Modeling and Estimation

The current in the motor windings generates a magnetic field in the air gap of the motor. This field interacts with the magnetic field from the magnets of the rotor and produces the output torque. The torque constant is the ratio of the

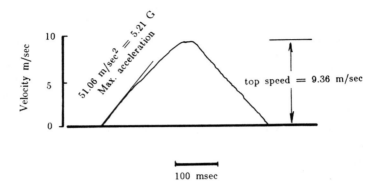

Figure 8-5:   Velocity profile of base motor

Arm configuration $\alpha_2 - \alpha_1$ Deg.		Oscillation Amplitude Ratio (Db) of Passive Axis to Active Axis	
		0.5 Hz	5 Hz
Effect of motor 2 on motor 1	95	-48	-31
	110	-57	-40
	130	-53	-39
	140	-52	-32
Effect of motor 2 on base motor	95	-57	-34
	110	-59	-38
Effect of motor 1 on base motor	95	-59	-49

Table 8-2: Experimental data on interactions among actuators

output torque to the current. The magnetic field changes with the rotor position and causes variations in the torque constant. This results in torque ripple. Although the motors used for direct-drive arms are designed so that the torque constant is as uniform as possible for all the rotor positions, there still exists a significant torque ripple.

The torque generated is then transmitted to the joint axis through the motor shaft and the joint housing. In the case of a traditional joint mechanism, the torque is transmitted and amplified through a reducer, which causes the large torque disturbances produced by friction, backlash and dynamic deflections. The elimination of the gearing in the direct-drive arm greatly reduces these disturbances.

The motors, described previously, are three-phase variable frequency synchronous motors developed for the MIT direct-drive arm [Asada and Youcef-Toumi 83a, Davis and Chen 84]. The motors are capable of exerting large torque so that the gearing can be completely eliminated. The motor is also designed to minimize the torque ripple. In the sinusoidal torque control generation scheme [Tal 81], the relationship between the output torque and the currents flowing into the three-phase windings was discussed in Chapter 2 and is given by

$$\tau = K_t \left[I_{am}\sin \Theta_{rot} + I_{bm}\sin (\Theta_{rot} + 120) + I_{cm}\sin (\Theta_{rot} + 240) \right] \tag{8.2}$$

where I_{am}, I_{bm} and I_{cm} are three phase currents, Θ_{rot} is the angular position of the motor rotor and K_t is the torque constant.

From the above relationship, it follows that the output torque can be estimated by measuring the three phase currents and the angular position of the motor rotor. The three phase currents can be measured by inserting sensing resistors in series with the phases of the motor, and the rotor position can be measured by a position transducer attached to the rotor shaft. In the synchronous motor used, both the current and position information is already available, so that no additional sensors are necessary for implementing the torque measurement.

The joint torque estimator discussed in the previous sections was implemented and tested in terms of static and dynamic performances, [Asada, Youcef-Toumi and Lim 84] . The three phase currents were monitored with 0.01 Ω sensing resistors through circuits which isolate the measured signals from the high voltage power lines of the drive amplifier. The computation of the output torque was accomplished with a computer by recording the measured currents and the rotor position information provided by a synchro resolver, refer to Appendix B.

Figure 8-6 shows the results of an experiment using the current torque sensor in which applied and estimated torques were compared for the same rotor position. The estimated torques show no hysteresis and a linear relationship with the applied torques except around the origin. This discrepency is mainly due to the nonlinearity of the isolation circuitry. The sensitivities were 0.223 *V/Nm* and 0.260 *V/Nm* for applied torques in the clockwise and counterclockwise directions respectively. The difference in slopes is caused by a slight misadjustment of the resolver. The currents in each phase were also tested for repeatability and consistency. Various conditions were used for this test including (i) loading and unloading of the applied torques, (ii) moving away and back to the test position, (iii) and turning the system on and off. The repeatability in the current readings is due to the closed loop current controller in the amplifier that regulates current in each phase.

The accuracy of this method depends not only on the accuracy of the current and position tranducers but also on the variation of motor constant k_t. Since the magnetic field in the air gap varies slightly with the rotor position, the torque estimation based on Equ. (8.2) will also vary with the rotor position. Distortions of the sinusoidal motor phase currents I_{am}, I_{bm} and I_{cm} cause additional errors in the torque estimated. The experimental results in Figure 8-7 show the effects of these errors on the estimated torque. A constant torque was applied externally to the motor shaft for various rotor positions. The error in the computed torque was ± 10 % of the mean value. However, the estimated torque has a clear pattern where two main frequencies are dominant. The two lowest frequencies correspond with the number of rotor poles and the number of stator slots respectively. This error can be reduced by modifying equation (1) so that the errors due to the rotor and stator slot locations are compensated for. With this modification, the torque measurement error was reduced to ± 6 % of the mean value.

8.5.2 Torque Control

[Asada and Lim 85] evaluated the dynamic response of the torque control system. The step response showed a delay time and a settling time of 6 and 60 msec respectively. It was also found that the bandwidth of the torque control system was about 30 Hz. Thus a fast and accurate torque control system has been achieved. Article 3 of Part IV discusses these issues in detail.

8.6 Conclusion

The control performances of the developed prototype direct-drive robot were considered. The manipulator, equipped with high torque brushless motors and a parallel drive five-bar-link and six-bar-link mechanism with a decoupled and nearly invariant arm inertia tensor, exhibits excellent performance. The delay time in the response is 27 ms, and the top speed and acceleration are on the order of 10 m/s and 5 G respectively. In addition, high servo stiffnesses were achieved. These characteristics are each an order of magnitude better than those of conventional robots. Also, the interactions between joints were also reduced to less than -30 Db's for all arm configurations. In addition, the torque developed by the direct-drive motor was estimated through the measurement of phase currents. Using a simple correction function, that depends on the pole and slot ripples, the estimated torque was found to be within ± 6 % of the mean.

154

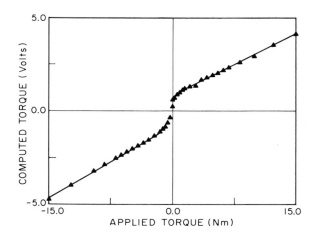

Figure 8-6: Calibration curve of the current torque estimator

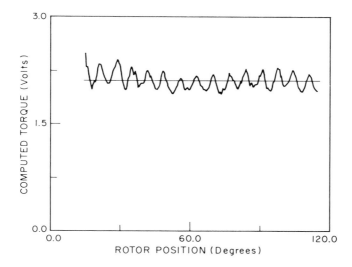

Figure 8-7: Experimental data in computed torque
variations without correction

Part IV

SUPPORTING ARTICLES

Chapter 9

Control of a Direct-Drive Arm

Reprinted from September 1983, Vol. 105, Journal of Dynamic Systems, Measurement, and Control[5]

Haruhiko Asada[7] Assistant Professor, Department of Mechanical Engineering, Massachusetts Institute of Technology, Cambridge, Mass. 02139 Assoc. Mem. ASME

T. Kanade Senior Research Scientist, Robotics Institute, Carnegie-Mellon University, Pittsburg, Pa. 15213

I. Takeyama[6] Kubota Ltd., Osaka, Japan

[5]This work was supported by The Office of Naval Research (Contract No. N00014-81-K-0503).

[6]Formerly with Canegie-Mellon University where this work was done. Contributed by the Dynamics Systems and Control Division for publication in the JOURNAL OF DYNAMIC SYSTEMS, MEASUREMENT, AND CONTROL. Manuscript received by the Dynamic Systems and Control Division, May 24, 1982.

ABSTRACT

A direct-drive arm is a mechanical arm in which the shafts of articulated joints are directly coupled to the rotors of motors with high torque. Since the arm does not contain transmission mechanisms between the motors and their loads, the drive system has no backlash, small friction, and high mechanical stiffness, all of which are desirable for fast, accurate, and versatile robots. First, the prototype robot is described, and basic feedback compensation is discussed. This compensation significantly reduces the effect of interactions among multiple joints and nonlinear forces. The experiments showed the excellent performance of the direct-drive arm in terms of speed and accuracy.

9.1 Introduction

As robots find more and more advanced applications, such as assembly in manufacturing, accurate, fast, and versatile manipulation becomes necessary. One of the difficulties in controlling mechanical arms is that they are highly nonlinear and involve coupling among the multiple links. Recent progress in the analysis of arm dynamics allows the real-time computation of full dynamics using efficient algorithms in a recursive Lagrangian formulation [Hollerbach 80], a Newton-euler formulation [Luh, Walker, and Paul 80], [Stepanenko and Vukobratovic 76], or a look-up technique [Horn and Raibert 78]. When the arm dynamics are identified accurately, the feedforward compensation of nonlinear and interactive torques combined with optimal regulators for linearized systems improves control performance greatly [Vukobratovic and Stokic 80] and guarantees a global stability over a wide range of arm configurations [Vukobratovic and Stokic 81].

A critical obstacle to use of dynamic models lies in the uncertainty of arm dynamics or the difficulty of identification. Even a single component in a drive system, such as a gear, a lead screw, a steel belt, or a servovalve and a pipe has complex and changeable characteristics in terms of friction, deflection, backlash, compressibility, and wear. When the arm dynamics are not well identified and subject to unknown payloads, one can use model-referenced adaptive control to maintain a uniform performance [Dubowsky and Des Forges 79]. Its extensive application was shown to allow decoupling of the arm dynamics in a Cartesian coordinate system [Takegaki and Arimoto 81], the reduction on computational burdens [Horowitz and Tomizuka 81], and high speed manipulation [Leborgne, Ibarra, and Espiau 81].

Alternatively, a rather straightforward way to achieve high-quality dynamic performance is to pursue the development of a new mechanism which can be modeled accurately with little difficulty. The obedient dynamical characteristics of the simplified arm may make it easy and effective to apply sophisticated control. A direct-drive arm is such a new mechanical arm which radically departs from the comventional arm mechanism. In a direct-drive arm, unlike a conventional mechanical arm that is driven through gears, chains, and lead screws, the joint axes are directly coupled to rotors of high-torque electric motors, and therefore no transmission mechanism is included between the motors and their loads. Because of this, the drive system has excellent features, no backlash, small friction, and high stiffness. The authors have developed the first prototype of the direct-drive arm with six degrees of freedom [Asada and Kanade 83]. The simple mechanism allows us to have the clear and precise model of the arm dynamics, which is of special importance not only for accurate positioning control but also for compensating interactive and nonlinear torques in high speed manipulation. This paper describes a characteristics analysis of the direct-drive arm and the design of a control system to ahcieve the excellent performance that the direct-drive arm potentially has.

9.2 Outline of the Direct-Drive Arm

The overall view of the developed direct-drive arm is shown in Photo 9-1 and its assembly drawing in Fig. 9-1. The arm has 6 degrees of freedom, all of which are articulated direct-drive joints. Beginning from the upper base frame, the first joint is a rotational joint about a vertical axis, and the second is a rotational joint about a horizontal axis. The third and fourth joints rotate the forearm about the center axis of the upper arm and about its perpendicular axis, respectively. The fifth and sixth joints perform a rotational and a bending motion of the wrist part. The total length of the arm is 1.7 m, and the movable part form joint 2 to the tip is 1 m. The movable range of joints 1 and 5 is 330 degrees. The remaining joints can move 180 degrees. The maximum payload is 6 kg including a gripper attached at the tip of the arm.

High performance d-c torque motors were used for the direct-drive arm. Photo 9-2 shows a bare motor at the first joint. The motor consists of a rotor, a

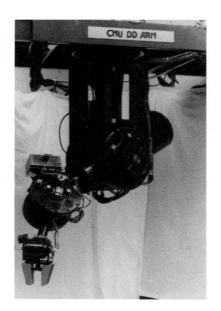

Photo 9-1:   Overall view of direct-drive arm

Photo 9-2: Bare motor at shoulder joint (joint 1).

stator, and a brush ring. As shown in Photo 9-2, each component of the motor is installed directly at the joint housing; the rotor on a hollow shaft, and the stator and the brush ring at a case. To develop a torque large enough to rotate the joint shaft directly, we selected motors with large diameters. The motor to drive joint 1 is 56 cm in diameter with 204 Nm peak torque. Joint 2 has two motors, one on each side of the upper arm. These motors are 30 cm in diameter with 136 Nm peak torque. It is required that the motors at joints 4, 5, and 6 have not only high torque but are also lightweight and compact, because heavy motors at these joints would apply a large load on upper joints. Therefore we used high performance torque motors with samarium cobalt magnets. The maximum magnetic energy product ot these magnets is 3 to 10 times larger than that of conventional ferrite or alnico magnets [Asada and Kanade 83]. The two samarium cobalt motors to drive joint 4 are 23 cm in diameter with 54 Nm peak torque, and the motors for the last two joints are 8 cm in diameter with 6.8 Nm peak torque.

An optical shaft encoder was installed at each joint to measure the joint angle and its angular velocity. We used precise encoders combined with accurate gears with 1 to 4 or 1 to 8 gear ratios. The resultant pulse density is 2^{16} pulses per revolution for the first four joints and 2^{15} pulses per revolution for the last two joints.

9.3 Mathematical Modeling and Identification

9.3.1 Kinematics and Dynamics

We modeled the kinematic structure of the arm according to the Denavit and Hartenberg convention [Denavit and Hartenberg 55]. The arm consists of 7 links numbered 0 to 6 form the base to the tip of the arm. Joint i is the joint that connects link i to link $i + 1$. Figure 9-2 shows the disjointed links of the direct-drive arm where the rotors and stators of motors are disassembled and attached to separate links. Each link has a coordinate frame fixed to the link, where the z axis points the direction of the joint axis. The geometry of each link is described by the three parameters listed in Table 9-1. Joint displacement is given by joint angle θ_i that is the angle between the x_{i-1} and x_i axes measured in a righthand sense about z_{i-1}. The three parameters listed in Table 9-1 and the above joint displacement completely describe the relation between any adjacent

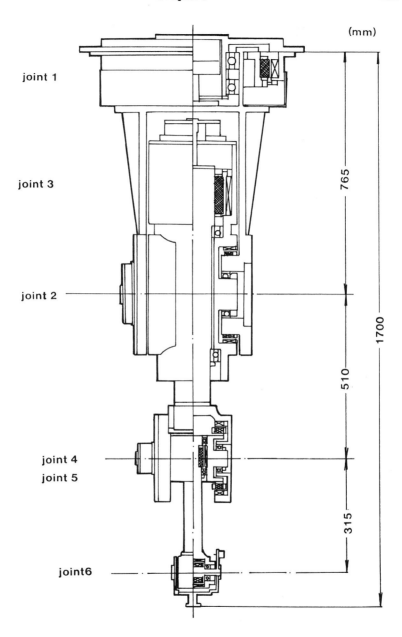

Figure 9-1: Drawing of direct-drive arm

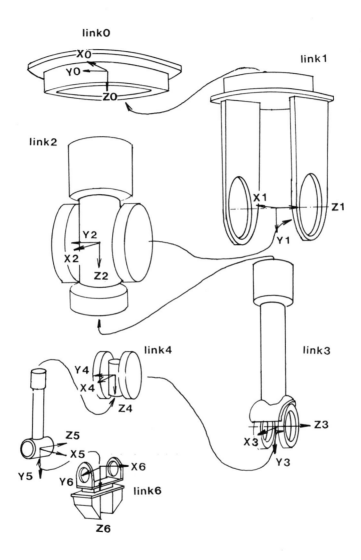

Figure 9-2: Disjointed links and local coordinate frames

a_i the distance between coordinate frames $i-1$ and i measured along x_i,

s_i the distance between x_{i-1} and x_i measured alond z_{i-1},

α_i the angle between the z_{i-1} and z_i axes measured in a right-hand sense about x_i,

joint number	a_i [m]	s_i [m]	α_i [deg]
1	0	0.765	180
2	0	0	-180
3	-0.035	0.510	180
4	0	0	-180
5	0	0.315	180
6	0	0	-180

Table 9-1: Description of arm structure in Denavit and Hartenberg convention

links and the total arm configuration.

The equation of motion of the arm is derived assuming that the arm consists of a series of rigid bodies. The characteristics of a single link are completely represented by mass, center of mass, and moment of inertia about this center. We computed these parameters for each link of the direct-drive arm form the detail drawings. Table 9-2 shows the result, where the center of mass and the moment of inertia are described in each link-coordinate frame.

Kinetic energy and potential energy stored into the arm are obtained by using the data listed in Tables 9-1 and 9-2. Let joint angles θ_i and joint torques τ_i be generalized coordinates and generalized forces, respectively, then the following equation of motion is derived from substituting the dinetic and potential energies into the Lagrange's equation of motion.

$$\tau_i = \sum J_{ij}\ddot{\theta}_j + \sum\sum b_{ijk}\dot{\theta}_j\dot{\theta}_k + f_{gi} + f_{di} \qquad (9.1)$$

where the first term on the right-hand side stands for inertial forces, the second term, consisting of products of angular velocities, stands for Coriolis and centrifugal forces, the third term represents gravity load, and the last term stands for the other disturbing torque such as friction and external force. The direct-drive joints have friction only at the bearings that support joint axes and at the brushes between the rotors and stators. These friction coefficients are negligibly small for most of the direct-drive joints. The parameters J_{ij}, b_{ijk}, and f_{gi} vary depending on the arm configuration, namely the functions of $\theta_1, ..., \theta_6$.

9.3.2 Drive Systems

Since the motor of a direct-drive joint is directly coupled to its joint axis, the driving torque about the axis is exactly the same as the torque developed by the motor, which is proportional to current I_i applied to the motor armature;

$$\tau_i = K_{ti}I_i \qquad (9.2)$$

where K_{ti} is the torque constant. The electric characteristics of the armature are given by

$$V_i = R_iI_i + E_i \qquad (9.3)$$

link number	mass (kg)	center of mass (m)	moment of inertia (kgm^2)		
1	95.99	0.000 -0.675 -0.015	33.724 0.000 0.000	0.000 2.879 0.609	0.000 0.609 33.056
2	82.61	0.000 0.010 -0.203	3.990 0.000 0.001	0.000 3.786 -0.154	0.001 -0.154 1.475
3	52.90	0.029 -0.524 -0.007	8.295 0.178 -0.012	0.178 0.425 0.195	-0.012 0.195 8.237
4	13.34	-0.001 0.024 0.002	0.150 0.000 -0.001	0.000 0.062 0.001	-0.001 0.001 0.150
5	4.84	0.002 -0.176 0.000	0.110 -0.002 0.000	-0.002 0.005 0.000	0.000 0.000 0.110
6	2.81	0.000 0.008 0.032	0.016 0.000 0.000	0.000 0.011 0.002	0.000 0.002 0.006

Table 9-2: Mass, center of mass, and moment of inertia

where V_i is applied voltage to the armature of the motor i, R_i is the resistance of the armature, and E_i is the back EMF. Since the inductance of the armature is small, it is neglected in the above equation. The back EMF is proportional to the angular velocity of the joint axis and is given by

$$E_i = K_{ti}\dot{\theta}_i \tag{9.4}$$

where the back EMF constant is the same as the torque constant in metric units. Let u_i and K_{ai} be the input voltage and the voltage gain of a servo amplifier, then the characteristics equation of the drive system is derived from equations (9.2), (9.3), and (9.4);

$$\frac{K_{ai}K_{ti}}{R_i} u_i = \tau_i + \frac{K_{ti}^2}{R_i}\dot{\theta}_i \tag{9.5}$$

Thus the drive system is characterized by the following parameters

$$K_{ai}^* = \frac{K_{ai}K_{ti}}{R_i}, \quad C_i = \frac{K_{ti}^2}{R_i} \tag{9.6}$$

where K_{ai}^* is torque gain between the input u_i and the exerted torque τ_i, and C_i represents the coeffieient of damping force inherent to the drive system. These parameters were experimentally determined. The torque gains of joints 1,4, and 6, which are joints at the shoulder, elbow, and wrist of the arm, were 15.88 Nm/volt, and 2.15 Nm/volt, respectively. The coefficients of damping force of these joints were 18.17 NmS/rad, 2.31 NmS/rad, and 0.148 NmS/rad, respectively.

9.3.3 Single-Link Model and Frequency Response

As a first step to investigate the characteristics of servomechanisms, we assume a simplified load for each actuator. Namely, we first neglected all the nonlinear effects such as Coriolis and centrifugal forces as well as gravity and friction. We also assumed that when joint i is investigated all the other joints are mechanically immobilized. The equation of motion of the arm is then: $\tau_i = J_{ii}\ddot{\theta}_i$, because $\theta_j = 0$ and $\dot{\theta}_j = 0$, for $j \neq i$. Figure 9-3 shows the block diagram of the single-link drive system. The blocks enclosed by a broken line

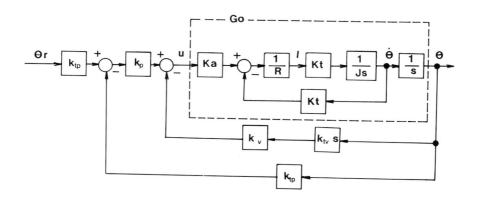

Figure 9-3: Block diagram of single-link control system

represent the control object including a servo amplifier, a motor, and the simplified load. The velocity feedback inside of the control object represents the back EMF of the motor. The transfer function of the control object is then given by

$$Go_i(s) = \frac{K_i}{s(T_i s + 1)} \quad , \quad K_i = \frac{K_{ai}{}^*}{C_i} \quad , \quad T_i = \frac{J_{ii}}{C_i} \tag{9.7}$$

We identified the single-link drive systems through experiments. Figures 9-4 and 9-5 show the results of frequency response for joints 1 and 4, which are joints at the shoulder and elbow of the arm. The inertia load of each joint varies depending on the arm configuration. In particular, the inertia load on joint 1 varies remarkably relative to joint 2. The experiments were carried out for three different arm configurations $\theta_2 = 0$, 45, and 90 deg. Since the phase curves do not exceed -180 degrees for these configurations, the control object can be identified as a second-order system whose order coincides with the theoretical model just derived. The solid curves in the figures show the models derived from the parameters listed in Tables 9-1 and 9-2. The good agreement between the models and the experimental data shows that the identified parameters were correct.

In general, the identification of the indirect-drive robots is more difficult because of the friction and backlash existing at the transmission mechanisms. Also, the deflection at the transmissions sometimes causes significantly large delay in higher frequency. The feature of the direct-drive robots is that it contains fewer uncertain factors and no higher-order delay. Hence the system can be identified accurately.

On the other hand, the direct-drive arm has the following problem. Let us evaluate the coefficients of damping force C_i relative to the inertia J_{ii}. From equation (9.7), the ratio of J_{ii} to C_i is the time constant of the control object. The time constants that can be determined from Figs. 9-4 and 9-5 are significantly large: $T_1 = 269$ ms for joint 1 in case of $\theta_2 = 45$ deg, and $T_4 = 269$ ms for joint 4. Namely, the direct-drive robot has large inertia loads relative to the damping characteristics inherent to the motors. Therefore we need to improve the damping characteristics for stabilizing the system.

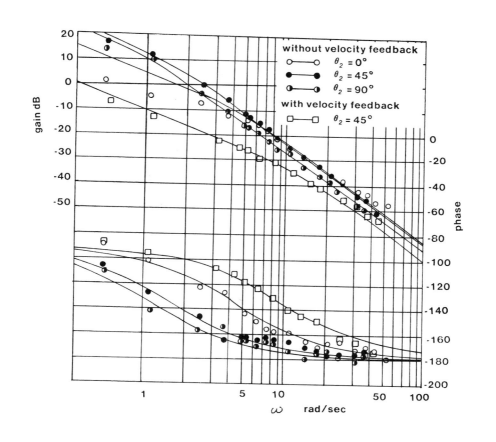

Figure 9-4: Frequency response of joint 1

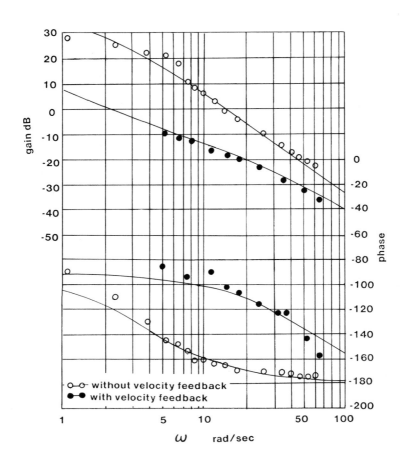

Figure 9-5: Frequency response of joint 4

9.4 Feedback Control

9.4.1 Velocity Feedback

In this section, we discuss velocity feedback to increase damping of the direct-drive joints. In the case of an indirect-drive joint, velocity is usually measured at the shaft of a motor before the speed is reduced by gears. However, it is rather difficult to do so in a direct-drive joint, because the speed of motor is a slow as the link motion. We employed high resolution shaft encoders and developed electric circuits to measure the slow speed movement of the direct-drive joint. The slowest speed that the developed system can detect was 2 deg/s for the first four joints and 4 deg/sec for the last two joints. The maximum speed, on the other hand, was 180 deg/sec and 360 deg/s, respectively. The resolution was 1/256 of the maximum speed in both cases

As shown in Fig. 9-3, we implemented the velocity feedback using the encoders, and evaluated the performance through experiments. When the velocity feedback gain Kv is increased, the damping characteristics are improved. However, if the gain is extremely large, it amplifies the error as well as the signal. When a joint rotates near the minimum detectable speed, the velocity signal varies frequently between zero and the minimum value. This varying velocity signal gives a large fluctuation of control torque and decreases control accuracy. We determined the maxumum allowable gain that does not cause such a vibration, by ovserving the motion in slow speed control.

Figures 9-4 and 9-5 also show frequency responses of joints 1 and 4, respectively, after the velocity feedback compensation was done using the maximum allowable gains. The phase curves show a noticeable phase lead about 50 to 60 degrees. By fitting theoretical curves to the experimental data, we obtained the time constants for the improved response. The time constants of the improved systems are 92 ms and 19 ms for joints 1 and 4, respectively. The velocity feedback compensation decreases the time constants 6 to 14 times smaller than those without it.

9.4.2 Gain Adjustment

Now we proceed to the gain adjustment for the improved systems. Figure 9-3 includes a position feedback loop, where k_p is the feedback gain to be adjusted. Since overshoot is usually undesirable in the control of mechanical arms, we adjust the position feedback gain Kp so that the damping factor is between 0.9 and 1. Figure 9-6 shows the step response for joints 1 and 4: response (a) is overdamped, response (b) is underdamped, and response (c) is critically damped. The responses for the three joints are recorded in the same time scale. The response of joint 1, which has a large inertia load, is relatively slow, while joints 4 and 6 have very fast responses. To evaluate the transient response we used the 50 percent delay time Td and the 5 percent settling time Ts. The delay times of joints 4 and 6 were only 57 ms and 82 ms, respectively. They show that the developed direct-drive arm has excellent dynamics. Even joint 1 has a 365 ms delay time which is fast enough for most applications.

When a large step input is aplied to one of the joints, the joint may be accelerated to a fast top speed for a long distance motion. However, if the joint motion is accelerated to an excessively fast speed, it is dangerous and is not desired in some applications. We modified the feedback controller so that, if the speed exceeds an allowable operating range, the velocity feedback gain is increased to several times larger than the normal operation range. Figure 9-7 shows the experiments of transient response for a large step input. We determined the maximum speed for joints 1 to 4 was 180 deg/sec, and for joints 5 and 6, 360 deg/s. Both the joints in the figure, 4 and 6, are accelerated rapidly to the maximum speeds and settle to the reference input smoothly without overshooting. Thus the fast and stable response was achieved.

9.5 Feedforward Control

9.5.1 Control Scheme

As we analyzed in the previous section, the arm's behaviour is complicated in multiple-degree-of-freedom motion. In this section, we discuss the compensation for interactive torques among multiple links and nonlinear torques such as Coriolis, centrifugal, and gravity torques. Feedforward control effectively compensates all the predictable motions, as long as the characteristics of the arm are identified accurately. The direct-drive arm has the advantage that the simple structure allows us to have the accurate model of the control object.

By solving the equation of motion inversely, we can compute the torques to drive the arm along a specified trajectory [Hollerbach 80], [Luh, Walker, and Paul 80]. Let $\theta_{r1}, \theta_{r2}, ..., \theta_{r6}$ be a trajectory of joint angles. If the trajectory is

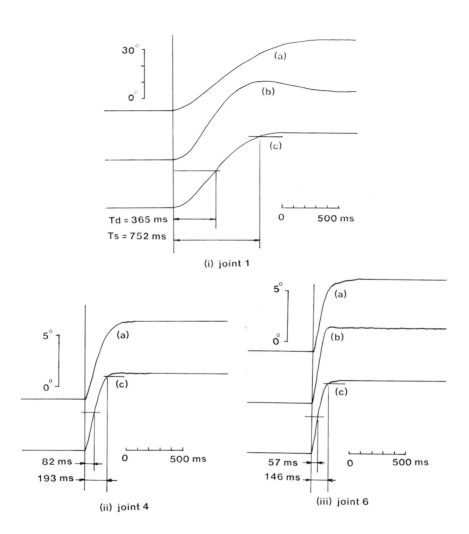

Figure 9-6: Experiments of step response

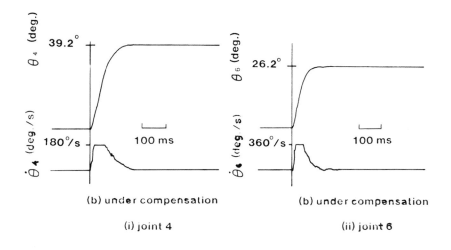

Figure 9-7: Response to large step input

smooth enough to differentiate up to the second order with respect to time, the
torques required to trace the trajectory with the specified speed and
acceleration, θ_{ri}, θ_{ri}, are derived from equation (9.1)

$$\tau_{ri} = \sum J_{ij}(\theta_r)\ddot{\theta}_{rj} + \sum b_{ijk}(\theta_r)\dot{\theta}_{rj}\dot{\theta}_{rk} + f_{gi}(\theta_r) \qquad (9.8)$$

where J_{ij}, b_{ijk}, and f_{gi} are functions of θ_{r1}, θ_{r2}, ..., θ_{r6}. If the identification of
the arm is perfect and no disturbing torque is applied to it, the arm can move
along the specified trajectory with the pre-computed torques. However, as the
arm travels for a long time, unavoidable errors are accumulated, even if the
identification error and disturbances are small. Since coefficients involved in
equation (9.8) are valid only when the arm configuration is in the vicinity of the
predicted state, the computed torques do not make sense as errors are
accumulated and the actual position of the arm diverges from the specified
trajectory. Therefore we need to keep the state of the arm close to the reference

trajectory. The feedback controller designed in the previous sections provides a continuous error correction of joint angles from the specified trajectory. We extended it to a controller that can correct the error of angular velocities from their references as well as the positional errors. By combining the feedforward control with the feedback control, we expected that the former would provide torques to lead the arm to a given trajectory with no delay and that the latter would provide the fine error correction to keep the state of the arm close to the reference. Thus the total torque applied to joint i is given by

$$\tau_i = \tau_{ri}(\theta_r) + kp_i^*(\theta_{ri} - \theta_i) + Kv_i^*(\dot{\theta}_{ri} - \dot{\theta}_i) \qquad (9.9)$$

where Kp_i^* is forward-path gain from the position reference to the torque of motor, and Kv_i^* is the resultant velocity feedback gain including the inherent damping due to the back EMF of motor and the artificial damping through velocity feedback;

$$Kp_i^* = K_{ai}^* Kp_i \quad , \quad Kv_i^* = C_i + K_{ai}^* Kv_i \qquad (9.10)$$

The second term and a part of the third term in equation (9.9) were already implemented in the feedback controller previously designed. Therefore, we provided the torques $\tau_{ri}(\theta_r)$ and $Kv_i^* \theta_{ri}$ through a computer that solves the inverse problem or arm dynamics.

9.5.2 Experiments

Figure 9-8 shows the experimental results of feedforward compensation for joint 4, where the sinusoidal inputs drawn by dash-and-dot lines were given to the system as reference trajectories and its responses after settling into steady oscillations were recorded. Curve (a) shows the case with no compensation, where significantly large offset and phase lag were observed as well as the reduction of amplitude. Curve (b) shows the case with the compensation of gravity torque, where the offset vanished and the amplitude was inlarged. In case of (c) where the damping torque, $kv_i^* \theta_{ri}$, as well as the gravity torque were compensated, a remarkable improvement of phase-lag can be seen. When all the arm dynamics were taken into account, the resultant response, curve (d), shows excellent agreement with the reference trajectory. We observed the excellent agreements with sinusoidal inputs over wide ranges of frequency and amplitude when the driving torque, angular velocity, and joint angle did not exceed their limits.

Figure 9-9 shows the responses of joints 4 and 6 for sinusoidal reference trajectories. When no feedforward compensation was applied, a noticeable interaction from joint 6 to joint 4 was observed. After the full dynamics of the two joints were compensated through the feedforward control, no significant interaction between them was observed and both trajectories showed excellent agreements with the regerences.

9.6 Evaluation of Steady-State Characteristics

9.6.1 Positioning Accuracy

In this section we evaluate the developed arm with respect to steady- state errors. Figure 9-10 summarizes the experiment of positioning accuracy, where the histogram of steady-state positioning errors for a step response is shown. Each histogram is obtained by more than 200 trials of the step response from the same point to the same destination. The horizontal line in each figure indicates the error from the destination (0 degree). Means and standard deviations were computed for each joint. To improve positioning accuracy, we used phase-lag compensators which increase loop gains 10 times larger in lower frequency. While joint 1, in figure (a), had a large offset 2.208 deg under no compensation, it was reduced to −0.287 deg, which is a reasonable error comparable to that of indirect-drive arms. However, the standard deviations indicated in the figure are very small; especially when the phase lag compesation was used, the deviation was only 0.019 def. The other joints, joints 4 and 6, have excellent positioning performance. The small standard deviations, 0.005 deg for joint 4 and 0.003 deg for joint 6, show that the direct-drive arm has a great advantage in terms of accuracy as well as speed. One of the reasons why the direct-drive arm shows the excellent repeatability is that the arm does not contain uncertain factors such as large friction at gears and deflection in chains and other flecible components.

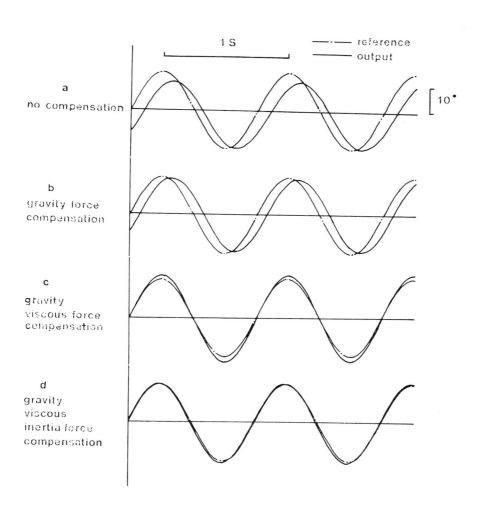

Figure 9-8: Effect of feedforward compensation for joint 4

Direct-Drive Robots

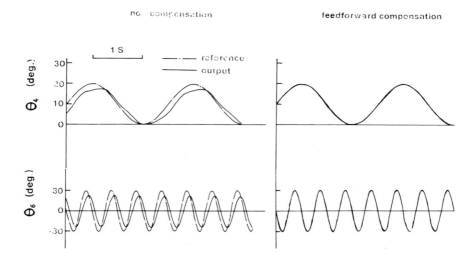

Figure 9-9: Effect of feedforward compensation for multiple joints
(joints 4 and 6)

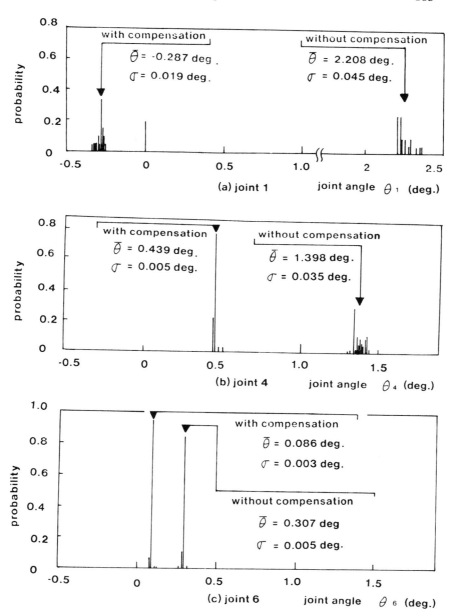

Figure 9-10: Experiments of positioning accuracy

9.6.2 Servo Stiffness

Although the direct-drive arm has fewer internal disturbances than indirect-drive arms, it is subject to external disturbances in actual operation. For example, the arm mechanically interacts with environments during manufacturing operations, or the arm grasps an unknown payload. Since these loads are not predictable in most cases, we could not compensate for them through the feedforward control discussed in the previous section.

Now we evaluate the sensitivity of the developed arm against external disturbances. Assuming that a disturbing torque Nd is applied to a joint axis, the steady-state error about this joint is $E = Nd/Kp^*$. To evaluate E we computed the deflection due to the load applied at the tip of each link. Suppose the link length is l and the disturbing force Fd is applied at the tip. Then $Fd = l \cdot Nd$, and the resultant deflection d at the tip of the link is $d = lE = l^2 Fd/Kp^*$, The servo stiffness Ks of a single-link drive system is defined by the ratio of applied force Fd to the deflection d [Paul 81].

$$Ks = \frac{Fd}{d} = \frac{kp^*}{l^2} \qquad (9.11)$$

The servo stiffness for joints 1, 4, and 6 were 3.2 N/mm, 10.6 N/mm, and 28.8 N/mm, respectively, under the phase-lag compensation, which were comparable to the stiffness of the Stanford Manipulator [Paul 81].

9.7 Conclusion

This paper has presented the development of a direct-drive arm and the evaluation of its control performance. After describing the outline of the developed direct-drive arm, we derived the mathematical model of the arm. The elimination of factors which are uncertain and hard to identify, such as friction, backlash, and deflection, made it possible to develop a precise model of arm dynamics and to employ it in arm control. At the same time, the modeling enabled us to extract important issues in controlling the direct-drive arm; poor damping characteristics and significance of interactive and nonlinear forces in arm dynamics. The experiments of control of the direct-drive arm have demonstrated the solutions of the control issued. First, it was shown that sufficient damping can be provided by velocity feedback using high-precision shaft encoders. Without overshooting, the arm responded to step inputs within 60 ms to 365 ms delay time and less than 180 deg/s or 360 deg/s maximum speed. Second, the experiment was performed on feedforward compensation of arm dynamics. A remarkable improvement in dynamic performance was observed. The significance of this experiment is that we have demonstrated usefulness of feedforward compensation by being able to model the arm dynamics precisely.

Chapter 10

A Linkage Design for Direct-Drive Robot Arms

Haruhiko Asada   Associate Professor

Il Hwan Ro   Graduate Student

Department of Mechanical Engineering, Laboratory for Manufacturing and productivity, Massachusetts Institute of Technology, Cambridge, Mass. 02139

Abstract

A new approach to manipulator link mechanism design is presented. Given torque-speed characteristics of actuators and an arm link mechanism, the resultant force-speed characteristics at the tip of the arm are analyzed. The link mechanism is optimized so that the end point speed and force are within appropriate ranges. This method is applied to a two-degree-of-freedom direct-drive arm. Direct-drive arms, in general, tend to have excessively fast operating ranges, whereas output forces are extremely small. A closed-loop five-bar-link mechanism is applied to the direct-drive arm, and the link dimensions are optimized in order to acheive appropriate force and speed ranges at the arm tip without using reducers.

10.1 Introduction

Servomotors, in general, are operated in high speed to exert large volumetric output power. Applying servomotors to manipulator drives, one needs gearings in order to match the motor characteristics with their load conditions. Since angular velocities of manipulator joints are much slower than that of servomotors, gearings used for electromechanical manipulators generally need large gear reduction ratios. Having such reducers, however, causes difficult problems due to backlash, friction, and poor mechanical stiffness, which degrade control performance [Asada and Kanade 83].

An alternative approach to manipulator drives is to use high torque, low speed motors and to replace or completely eliminate reducers with high gear ratios. Recent development of high torque motors, such as samarium-cobalt magnet brushless motors [Asada and Youcef-Toumi 83a], allowed us to build a direct-drive robot arm, which has demonstrated excellent features of this approach [Asada, Kanade and Takeyama 83], [Asada and Youcef-Toumi 84]. Nevertheless, the efficiency of motor power is still considerably low, because the normal operating speed of the arm is much lower than the speed in which the motors can exert power, efficiently. Motors with still higher torques and slower speeds are necessary to meet the operating condition, or some other means to reduce speeds and amplify torques are indispensable.

In this paper, an arm linkage design for a direct-drive robot is discussed in order to achieve appropriate speed reduction at the tip of the arm. The arm linkage as a whole is regarded as a transmission mechanism to transform the torque-speed characteristics of the motors. This would compensate for the mismatch between the motors and their loads and allow us to use direct-drive motors in an optimal manner without using conventional reducers. First, the force-speed characteristics in terms of the end point motion are analyzed, and their geometric representation is presented. This design tool is applied to an arm link mechanism for a direct-drive robot. The link mechanism is optimized and compared with conventional mechanisms for evaluation.

10.2 Force-Speed Characteristics

The torque-speed characteristics of a DC torque motor are given by a linear equation. Let V, τ, and ω be applied voltage to armature, output torque, and angular velocity, respectively, then the torque-speed characteristics are

$$\omega = \frac{1}{k_t} V - \frac{R}{k_t^2} \tau \qquad (10.1)$$

where R is the armature resistance and k_t is the torque constant, which is the same as the back emf constant in metric scale. For a given applied voltage V, the maximum torque and the maximum angular velocity are $\tau_{max} = k_t V/R$ and $\omega_{max} = V/k_t$, respectively. In order to use the motor for a direct-drive arm, τ_{max} must be sufficiently large. It is also desired that ω_{max} is on the same order as the maximum required speed of manipulation tasks. In general, output torques of direct-drive motors are rather small for driving manipulator arms, whereas maximum speeds are excessively large [Asada and Kanade 83], [Asada, Kanade and Takeyama 83], [Asada and Youcef-Toumi 84]. To describe how the motor characteristics are suited for the direct drive, let us evaluate the ratio of the maximum torque to the maximum speed:

$$\frac{\tau_{max}}{\omega_{max}} = \frac{k_t^2}{R} \qquad (10.2)$$

For efficient use, the ratio should be close to the one of the maximum torque and speed required for the tasks. The ratio of direct-drive motors, however, are generally small. The above ratio is independent from the applied voltage and is determined by the negative slope of the torque-speed characteristics in equation

(10.1). The square root of the ratio is known as the motor constant k_m that also represents the ratio of output torque to the square root ot power P dissipated in the windings [Electro-Craft 80].

$$k_m = \frac{k_t}{\sqrt{R}} = \frac{\tau}{\sqrt{P}} \qquad (10.3)$$

As the motor constant becomes larger, the motor can exert a larger torque with less dissipated power.

Specifications of the maximum torque and speed are originally given in terms of the arm tip motion in cartesian coordinates. From the torque-speed characteristics of all the motors driving an arm link mechanism, let us derive the resultant force-speed characteristics at the end point. For a system with n torque motors, equation (10.1) is rewritten in a vector form:

$$\omega = AV - B\tau \qquad (10.4)$$

where τ and ω are $n \times 1$ vectors representing output torques and angular speeds of n actuators in the system, respectively, A and B are $n \times n$ diagonal matrices whose ith entries are $1/k_{ti}$ and $1/k_{mi}^2$, respectively, and V is a $n \times 1$ vector representing applied voltages to n motors. End point velocities resulting from joint motions are given by using the Jacobian matrix J associated with cartesian coordinates of the end point and joint displacements.

$$\dot{X} = J\omega \qquad (10.5)$$

where $\dot{X}$ is a $n \times 1$ vector representing the end point velocities. End point forces, on the other hand, have the following relationship with the corresponding joint torques.

$$\tau = J^T F \qquad (10.6)$$

where J^T is the transpose of J and F is the end point forces. Substituting equations (10.4) and (10.6) into equation (10.5) the end point force-velocity characteristics are given by

$$\dot{X} = CV - LF \tag{10.7}$$

where

$$C = JA \quad \text{and} \quad L = JBJ^T$$

Equation (10.7) describes the force-velocity characteristics of the end point in terms of matrix L, the impedence matrix which relates the end point velocity to the output force vector [Asada and Youcef-Toumi 83a], [Asada and Youcef-Toumi 83b]. The impedence L is also a generalization of motor constants to a multi-degree-of-freedom system. The reciprocal of the squared motor constants are reflected to the end point of an arm link mechanism.

It is shown that the matrix L is a positive definite matrix for which there exist positive and real eigenvalues. Rewriting equation (10.7) using these eigenvalues, we obtain

$$\dot{q} = \begin{bmatrix} \dot{q}_1 \\ \vdots \\ \dot{q}_n \end{bmatrix} \tag{10.8}$$

$$\dot{q} = \begin{bmatrix} u_1 \\ \vdots \\ u_n \end{bmatrix} - \begin{bmatrix} \lambda_1 & & 0 \\ & \ddots & \\ 0 & & \lambda_n \end{bmatrix} \begin{bmatrix} f_1 \\ \vdots \\ f_n \end{bmatrix}$$

where f_i and $\dot{q}_i$ represent the force and velocity of the end point along the ith principal direction, repsectively. The column vector $U = (u_1 ... u_n)^T$ is related to matrix C as

$$U = T^T C V$$

where matrix T is the $n \times n$ orthonormal matrix associated with the principal transformation:

$$\dot{X} = T\dot{q} \tag{10.9}$$

Figure 10-1 shows the force-velocity characteristics at a given end point along the ith principle direction. We see from the figure that the eigenvalue defines the slope of the force-velocity curve and that the ratio of the maximum force and the maximum speed is also given by the eigenvalue. The force-velocity characteristics of the end point are represented geometrically by an ellipse whose length and direction of a major (minor) axis are determined by the minimum (maximum) eigenvalue and the corresponding eigenvector in a cartesian space [Asada and Youcef-Toumi 84]. As it can be seen form Fig 10-2, along the major (minor) axis, the impedance is the maximum (minimum). Namely, the arm can exert the largest (smallest) force, while the speed is the slowest (fastest). Thus the end point force-speed characteristics are easily comprehended with the geometric representation.

Note from equation (10.7) that the eigenvalues are the functions of the motor constants and the Jacobian matrix so that: $\lambda_i = \lambda_i(k_{m_1}...k_{m_n}, J)$. Previously, it was stated that the goal is to increase the end point forces and to decrease the end point speeds so that the force and speed ranges be close to those of normal manipulations. Namely, the goal is to reduce the eigenvalues, that is, to expand the principal axes of ellipses in their geometric representation. To achieve the goal, there are two possible ways. First one is to use motors with large motor constants. But, not only is this ineffective in terms of cost, it could also lead to unnecessary bulkiness in size [Asada and Kanade 83], [Asada and Youcef-Toumi 83a], [Asada and Youcef-Toumi 84], [Electro-Craft 80]. The other solution is to modify the arm link mechanism so that the associated Jacobian matrix gives larger ellipses. However, in order to have larger ellipses over the wide region of the workspace away from any singularity, a new design of arm mechanism is required. The following section introduces the proposed design of the arm mechanism based on five-bar link, closed kinematic chain

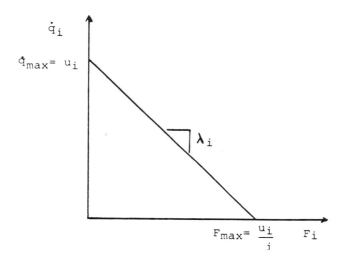

Figure 10-1: Force-speed characteristics

structure.

10.3 Kinematics of Five-Bar-Link Arm Mechanism

10.3.1 Kinematic Equations

In this section and the following section, the design method based on the force-speed characteristics of the arm's end point is applied to a two-degree-of-freedom planar mechanism for a direct-drive arm. The arm mechanism discussed in this section is a closed-loop five-bar-link mechanism shown in Fig. 10-3. There are two imput links that are driven by two independent motors, motors 1 and 2. The two motors are fixed to the base link, link 0, at point 0 and point A. The lengths of links 0,1, and 2 are denoted by l_0, l_1, and l_2, respectively. In addition to the base link and the two active links, there are two other links which are connected to links 1 and 2 and close the kinematic loop.

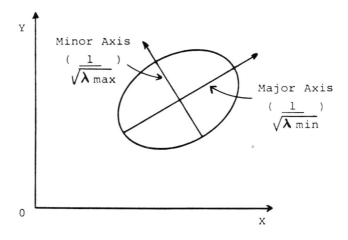

Figure 10-2: Geometric representation of force-speed characteristics

The lengths of the two links are denoted by l_3 and l_4. Point P represents the end point of the mechanism, which is fixed to link 4 at the distance of l_5 from point D with the angle of α from link 4, as shown in the figure. The coordinates of the end point are represented with reference to the coordinate axes fixed at point 0, where the X axis is directed along the base link. The angles of the input links are denoted by θ_1 and θ_2 measured from the X axis, and the angles of links 3 and 4 are by ϕ_1 and ϕ_2, respectively, measured from the X axis. There are totally 7 parameters that completely describe the arm mechanism and 4 joint angles including two independent input joints.

The end point coordinates, x and y, are given by

$$x = l_1\cos\theta_1 + l_5\cos(\phi_1 + \alpha) \tag{10.10}$$

$$y = l_1\sin\theta_1 + l_5\sin(\phi_1 + \alpha)$$

where ϕ_1 is a function of input angles, θ_1 and θ_2. The two input angles and the other two angles must satisfy the following relationships.

$$l_1\cos\theta_1 + l_4\cos\phi_1 = l_0 + l_2\cos\theta_2 + l_3\cos\phi_2 \tag{10.11}$$

$$l_1\sin\theta_1 + l_4\sin\phi_1 = l_2\sin\theta_2 + l_3\sin\phi_2$$

The above equation represents the constraint equation that must be satisfied to maintain the closed kinematic chain.

In order to obtain the force-speed characteristics of the end point, the Jacobian matrix associated with the relationship between the end point position and the input angles must be derived. Differentiating equations (10.10) and (10.11) in terms of θ_1, θ_2, ϕ_1, and ϕ_2, and then eliminating $d\phi_1$ and $d\phi_2$ from the differential equations, one obtains the relationship of infinitesimal displacements between $dX = (dx\ dy)^T$ and $d\theta = (d\theta_1\ d\phi_1)^T$.

$$dX = \mathbf{J}d\theta \tag{10.12}$$

where the Jacobian matrix $\mathbf{J}$ is given by

$$\mathbf{J} = \begin{bmatrix} J_{11} & J_{21} \\ J_{12} & J_{22} \end{bmatrix}$$

Where

$$J_{11} = -l_1\sin\theta_1 - \frac{l_1 l_5 \sin(\phi_1 + \alpha)\sin(\theta_1 - \phi_2)}{l_4 \sin(\phi_2 - \phi_1)} \qquad (10.13)$$

$$J_{12} = -\frac{l_2 l_5 \sin(\phi_1 + \alpha)\sin(\phi_2 - \theta_2)}{l_4 \sin(\phi_2 - \phi_1)}$$

$$J_{21} = l_1\cos\theta_1 + \frac{l_1 l_5 \cos(\phi_1 + \alpha)\sin(\theta_1 - \phi_2)}{l_4 \sin(\phi_2 - \phi_1)}$$

$$J_{22} = \frac{l_2 l_5 \cos(\phi_1 + \alpha)\sin(\phi_2 - \theta_2)}{l_4 \sin(\phi_2 - \phi_1)}$$

10.3.2 Singularities and Configuration Modes

This section discusses characteristics of the kinematic equations and the Jacobian matrix derived in the previous section. In general, the kinematic equations, (10.10) and (10.11), have four solutions that lead the end point to the same position. Figure 10-4 shows the arm configurations corresponding to the four solutions, where line parameter α is set to be zero. The links are coplanar and do not interact to each other. Each of the configurations is numbered 1 through 4 and is called mode 1 through mode 4.

Each mode associates with it a different Jacobian matrix, which relates to a different force-velocity characteristics. Having such multi-modes of configuration gives more versatility to the system compared to conventional systems, because the end point force-speed characteristics can be varied by changing the modes. Another interesting point is that each mode covers a different workspace. As shown in Fig. 10-5, the workspace generated by mode 1 (or 2) is different from the one for mode 3 (or 4). (The workspace was generated assuming a physical constraint, which is represented by crosshatched rectangular block.) The significance of this is that by utilizing such mode change, it is possible to assign each subworkspace at desired region of the work station.

The mode change is closely related to the singularities of the structure. This is because whenever mode change occurs, the linkage goes through a singular point. Thus, in this sense, the singular points define the boundaries

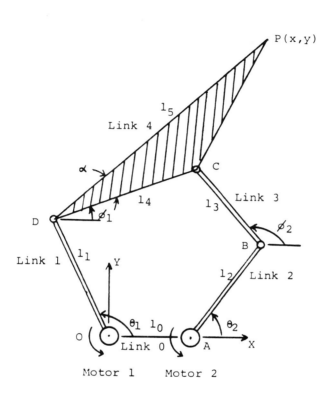

Figure 10-3: A five-bar-link planar mechanism

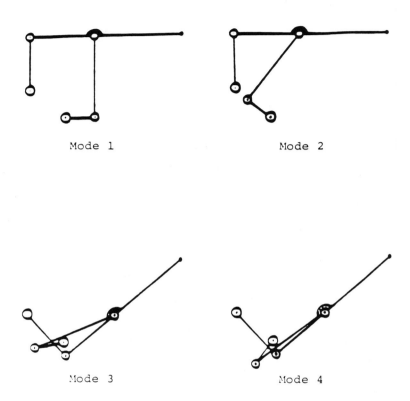

Figure 10-4: Four modes of configuration

between different modes of configuration. For the five-bar-link mechanism shown in Fig. 10-3, there are three different singularities. First, obtaining the determinant of the Jacobian matrix in equation (10.13), we get

$$|\mathbf{J}| = \frac{l_1 l_2 l_5}{l_4} \frac{\sin(\phi_1 + \alpha - \theta_1)\sin(\phi_2 - \theta_2)}{\sin(\phi_2 - \phi_1)} \tag{10.14}$$

In the above equation, there exist two conditions at which the determinant goes to zero, namely

i) $\sin(\phi_1 + \alpha - \theta_1) = 0$

ii) $\sin(\phi_2 - \theta_2) = 0$

From equation (10.13), it is found that the Jacobian matrix does not exist when:

iii) $\sin(\phi_2 - \phi_1) = 0$

This is another ill-condition which is inherent only of closed kinematic chains. Physically, when equation iii) is satisfied, points B, C, and D are aligned in Fig. 10-3. Therefore, the five-bar-linkage reduces to four-bar-linkage at that instant, and the rotation of the two input links become dependent as in a four-bar-link mechanism unless the displacements of motors at that postion are such that the input angle dependence is satisfied.

All three conditions define the boundaries in joint space. The first two define the boundaries between different modes of configuration whereas the last defines the boundary of physically reachable region. Figure 10-5 shows the plot of joint angle (θ_1 versus θ_2) for different modes and the boundaries defined by all three conditions. The two loci of dotted lines represent those joint angle positions at which the Jacobian does not exist. The first condition is represented by solid line, and the second by dashed line. The solid line defines the boundary between modes 1 and 2, whereas the dashed line defines the boundary between modes 3 and 4.

Also, the geometrical condition that is necessary for the smooth change of mode from one to another is

$$l_0 \geq |l_1 + l_4 - l_2 - l_3| \tag{10.15}$$

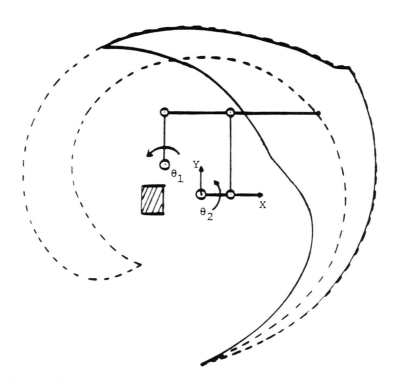

Figure 10-5: Workspace of all modes

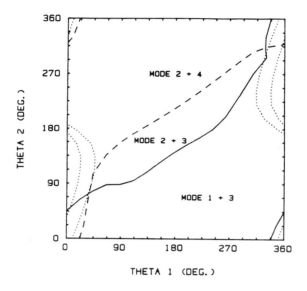

Figure 10-6: Plot of modes in joint space

The condition states that the length of the ground link must be long enough to allow both input links to fully rotate. Or in case of a four-bar-link mechanism, such condition is analogous to crank-crank mechanism, as opposed to crank-rocker mechanism.

10.4 Computations and Discussions

From the previous discussion, we found that there is a trade-off between obtaining the desired force-velocity characteristics and the wide workspace, that is, the larger the ellipses become, the smaller the reachable workspace. Figure 10-7 shows three different link mechanisms to illustrate this point. First, Fig. 10-7(a) shows the parallelogram five-bar-link mechanism covering a wide workspace, but its force-velocity characteristics are poor, as represented by small ellipses. On the other hand, Fig. 10-7(b) shows the case in which large ellipses represent significant improvement on force-velocity characteristics from the previous case, but the workspace is significantly reduced. This can be an excellent alternative for very fine assembly operation for which small workspace can be assigned. Finally, Fig. 10-7(c) shows the case where a desirable compromise has been reached. From the sizes of the ellipses, the force-velocity characteristics have been improved compared to the first case without sacrificing much of the workspace.

The optimization of force-velocity characteristics is limited to a region of the workspace. One way of optimizing is to define the region arbitrarily, and search for link combination which gives the desired end point characteristics. A program is written to generate the ellipses (axes lengths and the orientation of the principal axes with respect to inertial coordinate frame) for a given combination of normalized link parameters, which, in this case, are normalized with respect to output link. Then, the parameters are changed until the appropriate end point characteristics are obtained for that specified region of the workspace. Figure 10-8 shows the comparison between existing parallelogram mechanism and the new five-bar-link mechanism. After defining the region (the shape and the location are quite arbitrary), the ellipses are found for specified nodal points. From the figure, we observe a remarkable improvement by replacing the parallelogram mechanism with the new mechanism. Also, ellipses of the second case are much more isotropic than the first case, the ideal case of which becomes the circles, and the mechanism is equally force-efficient

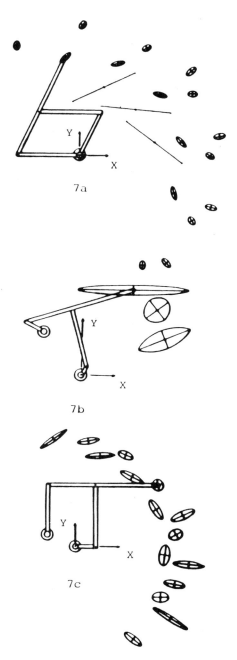

Figure 10-7: Force-speed characteristics

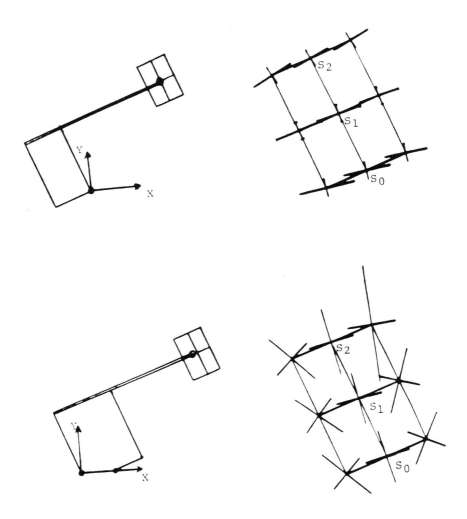

Figure 10-8: Force-speed characteristics of the mechanisms

to any arbitrary external forces applied at the tip of the arm.

Another aspect ot the new five-bar closed kinematic chain mechanism is that it has several modes of configuration. It has been discussed that each mode has a different Jacobian matrix associated with it and exhibits a different force-velocity characteristics. Figure 10-9 shows the comparative plots of the four modes of configuration on the basis of the lengths of major and minor axes of the ellipses at arbitrarily chosen points, in this case, at S_0, S_1, and S_2. The length of each major (minor) axis is normalized with respect to that of major (minor) axis of the corresponding ellipse of the parallelogram mechanism shown in Fig. 10-8(a).

$$r_1 = \frac{(\lambda_{max})10\text{-}8(a)}{(\lambda_{max})10\text{-}8(b)}, \quad r_2 = \frac{(\lambda_{min})10\text{-}8(a)}{(\lambda_{min})10\text{-}8(b)} \tag{10.16}$$

Fig. 10-9(a) is a plot of normalized lengths of major axes, r_1, for different modes whereas Fig. 10-9(b) is a plot of minor axes, r_2. From the plots, if the normalized length r_1 and/or r_2 of some modes approach infinity (or zero), that particular mode(s) is approaching a singularity. On the other hand, around the value of unity the particular mode(s) exhibitis no improvement of force-velocity characteristics with respect to that of the parallelogram mechanism.

It can be observed from the plots that modes 3 and 4 approach the singularity because r_1 for the modes approach infinite while the r_2 approach zero. Also, it can be observed from the plot that mode 1 exhibits more isotropic ellipses than mode 2 because, for mode 1, the normalized length r_2 is greater than r_1, which means the ellipses have become more circular, whereas from mode 2 the opposite case is observed.

10.5 Conclusion

This paper has presented the application of closed kinematic chain mechanism to improve the force-velocity characteristics of direct-drive robots, that is, to reduce the velocities and increase the forces at the tip of the arm. The program was written to generate the eigenvalues of the impedance matrix which governs the force-velocity characteristics. Among the various link combinations, one desirable link combination was chosen which gives the desired force-velocity characteristics within an arbitrarily specified region of

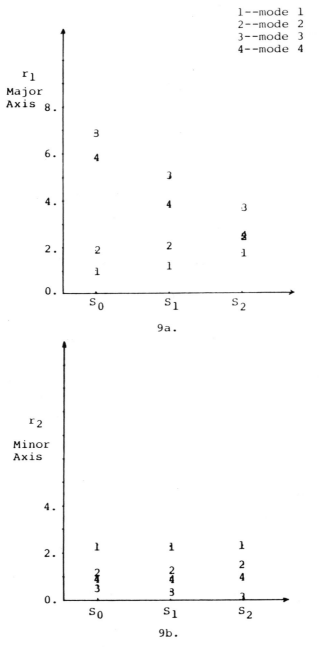

Figure 10-9: Plot of major and minor axes lengths of four modes

the workspace. Comparison has been made between the existing parallelogram mechanism and the new five-bar-link mechanism for evaluation. Also, the different modes of configuration has been studied on the basis of their effects on the end point force-velocity characteristics.

Chapter 11

Design of Joint Torque Sensors and Torque Feedback Control for Direct-Drive Arms

Haruhiko Asada Associate Professor

Swee-Keng Lim Graduate Student

Department of Mechanical Engineering, Laboratory for Manufacturing and Productivity, Massachusetts Institute of Technology, Cambridge, Mass. 02139

Abstract

In a direct-drive arm, motors are directly coupled to their arm linkages without reducers. This eliminates some defects of traditional gearing such as backlash, friction and deflection, and therefore enables the drive system to have improved control accuracy. The D.C. motors used for the direct-drive arm, however, have a significant amount of torque ripples and deadband, which limits the control accuracy, particularly for compliance and force control.

In this paper, a method for the torque feedback control of a direct-drive joint is presented. The motor output torque is directly measured by a strain gage sensor built into the motor. It is found that closed loop torque control can compensate for the complicated nonlinearities and improve the overall control accuracy.

First, the design of the torque sensor is presented. The effect of the structural vibration mode which results from the insertion of the flexible torque sensor is analysed, and a design method to place the natural frequency of this mode far above the operating frequencies is shown. Torque feedback control using the sensor is discussed. The sensor design is verified experimentally, and the control performance using direct torque feedback is evaluated.

11.1 Introduction

Manufacturing operations often require robots to mechanically interact with the environment. For these tasks, compliance motion control and force feedback control have been found to be useful by many researchers [Hanafusa and Asada 77, Hogan 85, Inoue 71, Mason 81, Paul and Shimano 76, Raibert and Craig 81, Salisbury 80, Takase 77, Weichbrodt and Beckman 78, Whitney 77]. The implementation of force control has been limited to simple operations. One of the reasons for this is that the hardware construction of presently available robots are not appropriate for compliance motion control and force control. Nonlinearities at reducers such as friction and backlash severely limit the performance of the compliance and force control. To reduce friction at reducers, Wu and Paul [Wu and Paul 80] have developed and implemented a joint torque sensor for a single joint manipulator. Luh et al [Luh, Fisher, and Paul 83] extended this work and implemented joint torque feedback on a Stanford manipulator. The system was shown to perform satisfactorily in reducing the effective friction in the harmonic drive of the robot arm. However, the backlash in the drive system produced a limit cycle.

The direct-drive arm, in which special high torque motors are directly coupled to their loads, has many advantages over its traditional counterparts as discussed by Asada et al [Asada and Kanade 83, Asada, Kanade and Takeyama 83, Asada and Ro 85, Asada and Youcef-Toumi 83a, Asada and Youcef-Toumi 84], particularly for compliance and force control. The elimination of reducers decreases friction remarkably. The complete elimination of backlash and the increase of mechanical stiffness also provide favorable conditions for force control, as well as for positioning control.

To fully exploit the advantages of the direct-drive arm and achieve high performance compliance and force control, it is necessary to further improve control performance of the motors and amplifiers. Recently, high torque, low speed motors appropriate for direct-drive arms became available. However, these motors still have significant nonlinearities such as torque ripples and deadband [Davis and Chen 84, Welburn 84]. These nonlinearities are generally highly complicated and difficult to identify accurately. As a result they degrade the control accuracy, particularly in the compliance and force control.

In this paper, a method for the control of direct-drive motors using torque feedback is described. The nonlinearities and uncertainties of the motors and amplifiers will be compensated for by the actual measurement of the output torques. This closed loop torque control will not only increase the torque control accuracy but also improve the positioning and trajectory control accuracy by reducing the effect of unmodeled nonlinearities and uncertainties. The actual torque measurement can also be used for advanced control applications. For example, mass properties of arm links can be accurately estimated by the torque measurement [Mukerjee and Ballard 85]. This estimation of mass properties will be useful to improve performance of adaptive control systems.

11.2 Torque Sensor Design

11.2.1 Design Issues

There are two ways of measuring motor torques used in robotics applications. One is to monitor motor currents and the other is to measure torsional deflection of the joint using strain gages. The former is based upon some known relationship between the input currents and the output motor torque. The accuracy of this method is limited by the accuracy of the model of the current-torque relationship. For direct-drive motors, particularly the motors used for the MIT Direct-Drive arm [Asada and Youcef-Toumi 83a], the current-torque relationship is highly nonlinear, hence it is difficult to identify accurately. This method is not, therefore, appropriate for accurate torque measurement and control. The second method provides a direct measurement of torque that does not depend on the accuracy of the electro-magnetic relationship.

Figure 1 shows a schematic construction of a joint torque sensor using strain gages. The torque transmitted from the motor rotor to the joint shaft is measured by the strain gages mounted on the flexible element. The sensitivity of the strain gage sensor depends on the stiffness of the flexible element connecting the motor to the load. For the same load, the lower the stiffness, the larger the strain becomes, hence the higher the sensitivity to the torque transmitted. To achieve accurate torque measurement, a lower stiffness member is desired. Having such a low stiffness member, however, conflicts with the other requirements, in particular accurate positioning control and system stability. Since high stiffness is one of the important features of the direct-drive arm, it is undesirable to insert such a flexible member into the drive mechanism. Thus, we need to solve the conflict between the high sensitivity required for torque measurement and the high stiffness required for positioning accuracy and stability. In the following section, we analyze the effect of the flexible member upon control performance and discuss a way of reducing the conflict in order to obtain high sensitivity torque measurement without the sacrifice of control performance.

11.2.2 Dynamic Analysis

Let us first consider the positioning error resulting from the deflection of the flexible member under a static load. When the displacement of the motor rotor, Θ_r, in Figure 1 is measured as the position feedback signal, an error results in the actual displacement of the arm link due to the direct loading upon the flexible member. If, instead of measuring the rotor displacement, the arm link displacement Θ_a is measured. The static error in the link position can be eliminated.

The dynamic deflection, however, is more complex. The insertion of the flexible member causes a torsional vibration mode between the motor rotor and the arm link. In order to preserve the dynamic performance of the original rigid system, the strain gage sensor must be designed so that this torsional mode becomes prominant only at frequencies much higher than the bandwidth of the original system.

Let k be the stiffness of the flexible member and J_r and J_a be the two moments of inertia of the mechanical system connected by the flexible member. The inertia of the rotor and the other members between the flexible member and the rotor are included in J_r, while J_a is the inertia of the arm link and the joint shaft. Note that, when a reducer is used, the rotor-side inertia J_r represents the equivalent inertia reflected to the joint shaft. The original rotor inertia is multiplied by the square of the gear ratio. The natural frequency of the system is then given

$$\omega_n{}^2 = \frac{k(J_r + J_a)}{J_r J_a} \tag{11.1}$$

Let us normalize the natural frequency ω_n with the natural frequency of the mass-spring system consisting of the arm link and the flexible member given by

$$\omega_a{}^2 = \frac{k}{J_a} \tag{11.2}$$

This represents the natural frequency when the rotor is immobilized. Let us also normalize the rotor-side inertia J_r by the arm-side inertia J_a, that is $r = J_r/J_a$, then the normalized natural frequency, ω_n/ω_a, is given by

$$\frac{\omega_n}{\omega_a} = \frac{1+r}{r} \tag{11.3}$$

The relationship between the normalized natural frequency ω_n/ω_a, and the normalized rotor inertia r is shown in Figure 2. The normalized natural frequency increases rapidly as the rotor-side inertia becomes smaller, while it approaches one as the rotor-side inertia becomes larger.

Figure 2 suggests an efficient way of solving the stability problem of the strain gage torque sensor, that is, the conflict between the high sensitivity requirement and the high stiffness requirement. The natural frequency can be increased without stiffening the flexible member of the sensor by making the rotor-side inertia J_r much smaller than the arm-side inertia J_a. For a small rotor-side inertia, the natural frequency becomes high enough to ignore its dynamic effect. The direct-drive arm has a significant advantage in satisfying this condition. The elimination of reducers makes the rotor-side inertia much smaller than that of the arm side, resulting in a very small normalized rotor inertia r, and hence a very high natural frequency. The rotor-side inertia J_r can be further reduced by locating the flexible member as close to the rotor body as possible. This suggests an appropriate sensor design for the dynamic motor torque measurement which will be described in the next chapter.

For conventional drive systems with reducers, this is difficult to design a sensor of this type that is both sensitive and stable. This is because the equivalent rotor-side inertia reflected to the joint shaft is of the same order as the arm side. Usually, the reducer is selected so that the rotor-side inertia is matched with the arm-side inertia; impedance matching. Under this condition, where $r = 1$ the natural frequency of the torsional mode becomes only 1.4 times as large as ω_a in Equ. ((11.1)). Therefore, we must make ω_a much higher than operation frequencies, otherwise the torsional mode becomes prominent. The use of a lower stiffness member is therefore strictly prohibited. Thus the high sensitivity requirement directly conflicts with the dynamics requirement.

11.3 Implementation and Experimentation for the Torque Sensor

In this section, the joint torque sensor is implemented and tested on the basis of the dynamic analysis in the previous section. The joint torque sensor is built into a direct-drive motor, Figure 3.

The motor is an 18-pole variable frequency synchronous motor with the stator windings and the samarium cobalt magnet rotor, while the stator external

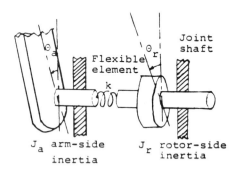

Figure 11-1: Schematic construction of arm link,
motor and torque sensor

diameter and rotor internal diameter are 254 mm and 139 mm respectively and
the peak torque of the motor is 130 Nm. The drive amplifier used is a 350 V, 4
kHz, PWM amplilfier that can drive up to 30 A output current.

The strain gage sensor is part of the hub supporting the motor rotor. The
sensor hub consists of an outer ring which is mounted to the motor rotor, an
inner ring which is coupled to the shaft, and three beams connecting the outer
and inner rings. All of them are made from a single piece of phosphor bronze,
which has low hysteresis. Each beam has a through-hole to increase the
sensitivity to the torsional stress. Strain gages are cemented on both sides of the
beam. This sensor beam structure has the advantage of high sensitivity as well
as high stiffness. Similar design has been used for a cutting force dynamometer
by Tani et al. [Tani, Hatamura, and Nagao 83].   To reduce noise, an
instrumentation amplifier is built in the motor so that wires between the strain
gages and the amplifier are shortened.

The torsional stiffness of the hub, k, is 1.5×10^4 Nm/rad. The sensitivity of
the torque sensor is 96.5 mV/Nm, which is sufficiently high compared to its low

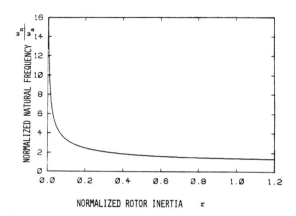

Figure 11-2:   Normalized natural frequency vs.
normalized rotor-side inertia

signal noise of about 5 mV. The arm-side inertia J_a is 0.24 kgm^2, while the rotor-side inertia J_r is only 0.01 kgm^2. From Equ. (11.1), the natural frequency of the system is then determined as 650 Hz, which is far above the normal operating frequency range of the system.

For this torque sensor, both the static and dynamic tests were carried out. Figure 4 shows the static test result. Torsional loads were applied to the joint axis in both the clockwise and counter-clockwise directions with the motor rotor rigidly fixed. The Figure shows the relationship between the applied torque and the torque measured by the torque sensor. The experiment was repeated for different rotor positions. The data showed excellent linearity and repeatability. The maximum measurement error was only about ±0.2 Nm, which is about the same as the friction torque in the bearings supporting the joint shaft.

To evaluate the effect of the flexible member upon the system dynamics the frequency response was measured and compared with the case when the flexible member was replaced by a rigid member. Figure 5 shows the comparison of the open-loop frequency responses between the two cases, where the outputs are joint velocities. It is found that both responses are identical over a wide range of frequency. However, at the high frequency range of about 105 Hz, a natural mode was observed for the flexible coupling case. This was determined to be the natural mode of the arm used. Another resonant mode was detected at about 590 Hz, which corresponds to the mode due to the flexible coupling between the shaft and the rotor by the sensor beams.

A closed-loop position control system was designed on the basis of the open-loop frequency responses. The bandwidth of the closed loop system is about 3 Hz. Which is an order of magnitude different from the two resonant frequencies. Due to the rapid decay of the loop gain at high frequencies, the torsional mode was not observed in the position control system. Thus, the insertion of a flexible member between the rotor and the joint axis does not impair position control.

Photo 11-1: Torque sensor built into direct-drive motor.
In the photo are shown: (1) stator, (2) rotor,
(3) shaft, (4) sensor beam, (5) strain gauges,
(6) amplifier circuit

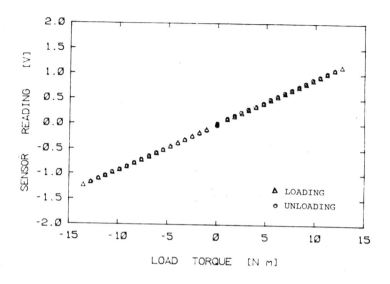

Figure 11-3: Static test of torque sensor

Direct-Drive Robots

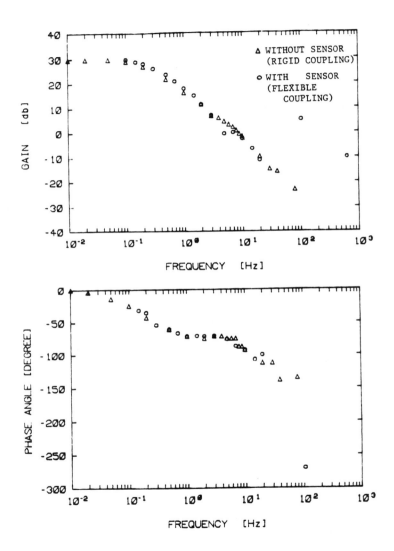

Figure 11-4: The open loop frequency response
of the motor

11.4 Torque Feedback Control

11.4.1 Controller Design

Using the joint torque sensor discussed in the previous sections, a torque feedback control system has been designed to improve the torque control accuracy. The goal was to show how the closed-loop torque control reduces errors resulting from the nonlinearities and the unmodeled dynamics of the motor and amplifier, e.g. torque ripples and deadband.

In compliance motion control, robot motion is constrained by contact with the environment. In this paper, we discuss dynamic behavior of a single-axis torque control where the arm link is rigidly constrained by the contact with a fixture. The dynamics of the control system in this case can be modeled as a first order delay due to the motor inductance and a second-order system due to the rotor-side inertia and the torsional stiffness. The open loop transfer function between the input command and the output torque is given by

$$G(s) = \frac{K_t}{s+\tau_e} \frac{1}{J_r s^2 + k}, \qquad (11.4)$$

where K_t and τ_e are the gain and electrical time constant of the motor-amplifier combination. The solid line in Figure 6 shows open-loop frequency responses of the motor output torque, when the arm link is rigidly fixed. The parameters involved in equation (11.4) are determined experimentally.

The torque control system has been designed on the basis of the frequency responses. To achieve high accuracy torque control, a high loop gain is desired. However, the resonant mode shown in Figure 6 is a major obstacle to the high loop-gain control. To increase the loop-gain without making the system unstable, the resonant mode must be at least a decade higher than the bandwidth of the system. To achieve this a small rotor-side inertia J_r is desirable. The advantage of the direct-drive arm is shown clearly in Figure 6. The simulated frequency responses of the conventional drive are shown by the broken line in the Figure. The system parameters are the same as the direct drive except that the rotor inertia is set equal to the arm-side inertia J_a, assuming that the impedance matching between J_a and J_r has been achieved. A highly underdamped resonant mode is observed at around 60 Hz for the conventional drive, while it is shifted to around 600 Hz in the case of direct drive. This difference significantly affects the design of the controller.

Now, we shall discuss the detail design of the torque control system for the direct-drive motor. As shown in Equ.((11.4)) as well as in the bode plot, the open-loop transfer function is a type 0 system, hence a steady-state error exists in the closed-loop system. In order to reduce the steady-state error, we use a lag-lead filter at low frequencies as shown by the chained line in Figure 6.

Since the bandwidth of the torque control system is much wider than the positioning control system, the highly underdamped resonant mode still affects the system stability even if the resonant frequency is in such a high frequency range. To remove the resonant mode, a notch filter was used as shown by the same bode plot in Figure 6. In addition, a low-passed filter is also used for further improving the dynamic response. The role of the low-pass filter is two fold. One is to reduce the electrical noise of the torque sensor. Although the sensor is carefully designed to have a low noise level, the torque control system is still disturbed by the noise since the system bandwidth is high for the torque control system. Another more important role of the low pass filter is to improve stability margins. The dynamic behaviour of the motor and amplifier is quite complicated and difficult to identify at such high frequencies, yet the stability margins of the torque control system are highly dependent upon the uncertain dynamics, particularly, in the frequency range from 100 to 500 Hz. The low pass filter makes the gain curve decays rapidly in this frequency range. The chained line in Figure 6 shows the resultant loop transfer function modified by the above compensator.

11.4.2 Experiments of Torque Feedback Control

In this section, the control performance of the torque feedback system is evaluated through experiments. Two experiments were carried out. In both experiments, the joint shaft was rigidly fixed to the base. The first experiment was to evaluate the control accuracy for different input commands at a fixed rotor position. The input command versus the output torque measured by the torque sensor is plotted in Figures 6.5 and 8. Figure 6.5 shows the output torque versus input command when no torque feedback control is employed. A distinct deadband of about ± 5 Nm was found as seen in the Figure. On the other hand, when feedback control is employed, the deadband was reduced as shown in Figure 8. In this figure, it is also clear that the hysteresis became much smaller than in the open loop case.

The second experiment was to examine the output torque as a function of the rotor position. As mentioned in the previous sections, the magnetic field in the motor air gap fluctuates significantly. This results in the variation of output torque as shown by triangles in Figure 9. In this figure a constant input command was given while the angular position was changed over 180 degrees. The data was taken at 5 degrees increment. When no torque feedback was employed prominent torque variation was observed. The fluctuation amounts to about 50 % of its mean magnitude. The output waveform resembles a sinusoidal signal with 5 peaks over half the circumference. This approximately corresponds to the number of pole-pairs in the motor in half the circumference. However, in the closed loop case, the fluctuation is significantly reduced. The standard deviations of torque fluctuation is only about 0.1 % of the mean magnitude.

Finally, dynamic response of the torque feedback control system was evaluated. Figure 10 shows the step response of the torque control system. The delay time and settling time are 6 and 60 ms respectively. From frequency response tests, it is found that the bandwidth of the system is about 30 Hz. Thus a fast and accurate torque control system was achieved.

11.5 Conclusion

A torque sensor for the direct-drive arm is designed. The effect of the structural mode introduced by the flexible element of the torque sensor is investigated. It was shown that for a direct-drive system where the effective rotor inertia is much smaller than the arm-side inertia, the use of a flexible element does not significantly degrade the control performance, as compared with conventional drive systems with reducers. With an appropriate sensor design, closed-loop position response was not affected by the presence of this structural mode.

A closed-loop torque control system was designed by using the torque sensor. This feedback control system significantly reduces the nonlinearities which exist in the original system. The torque fluctuation was reduced to about 0.1% of its mean value and the high bandwidth of 30 Hz was achieved as well.

Acknowledgement

The authors wish to express their gratitude to Drs K. Youcef-Toumi, H. Kazerooni for their advice and help in this work, and to H. West and N. Goldfine for their comments on this paper.

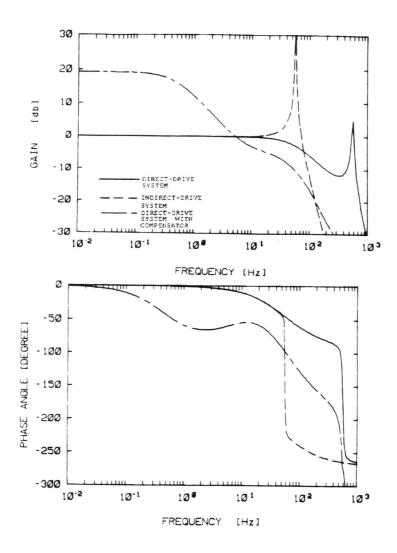

Figure 11-5: Bode plot of the static torque control system

Direct-Drive Robots

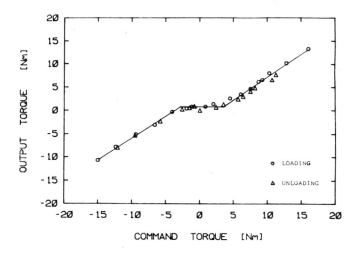

Figure 11-6:   Motor torque versus input voltage
                - open loop

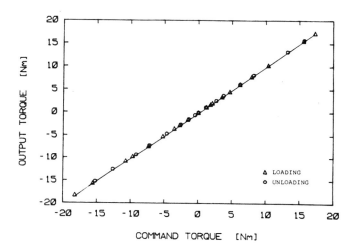

Figure 11-7: Motor torque versus input voltage
- closed loop

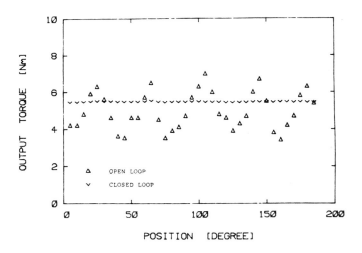

Figure 11-8:   Variation of torque
versus rotor position

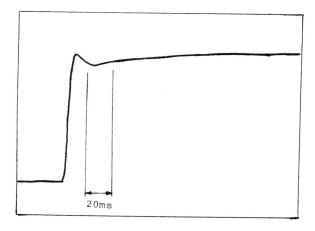

Figure 11-9:   Transient response of the torque loop

Part V

APPENDICES

Appendix A

Flow Correction in Current Reading

The measurement of current in each phase of the windings, shown in Figure A-1, is in error because of the sensory limitations. The current readings need to be corrected before their use. The correction is based on satisfying the continuity of flow, or Kirchoff's current law. The measured currents I_{am}, I_{bm} and I_{cm}, are corrected as follows [Paynter 82],

$$I_{acorr} = I_{am} - \frac{F}{F^*} |I_{am}|$$

$$I_{bcorr} = I_{bm} - \frac{F}{F^*} |I_{bm}|$$

$$I_{ccorr} = I_{cm} - \frac{F}{F^*} |I_{cm}| \qquad (A.1)$$

where

$$F = I_{am} + I_{bm} + I_{cm}$$

$$F^* = |I_{am}| + |I_{bm}| + |I_{cm}|$$

The above relations guarantee that continuity for the corrected currents is satisfied,

$$I_{acorr} + I_{bcorr} + I_{ccorr} = 0$$

234

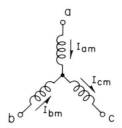

Figure A-1: 3 phase windings of the direct-drive motor

Appendix B

Torque Control Calibration and Software

B.1 Setup and Calibration

As presented in Chapter 8, the motor torque output is estimated from the phase currents. The relationship between the phase currents and torque output was given by Equ. (8.2),

$$\tau = K_t [I_{am}\sin(\theta_{rot}) + I_{bm}\sin(\theta_{rot} + 120) + I_{cm}\sin(\theta_{rot} + 240)] \qquad (B.1)$$

where the currents I_{am}, I_{bm} and I_{cm} are the measured currents in phases a, b and c respectively; K_t is the torque constant. The amplifiers for the direct-drive motors use the sinusoidal torque generation approach [Electro-Craft 80]. Thus the phase currents are sinusoidal functions of rotor position θ_{rot}. In practice, these are slightly distorted sinusoids. The weighting factors $\sin(\theta_{rot})$, $\sin(\theta_{rot} + 120)$ and $\sin(\theta_{rot} + 240)$, Equ. (B.1), are generated by software or by hardware using ROM's and multiplying D/A's (M/DA) as shown in Figure 7-2. The key issue in using Equ. (B.1) is to make sure that the sinusoidal measured current I_{am}, I_{bm} and I_{cm} are in phase with the generated sinusoids $\sin(\theta_{rot})$, $\sin(\theta_{rot} + 120)$ and $\sin(\theta_{rot} + 240)$ respectively. The torque constant K_t is determined experimentally from Equ. (B.1). The experimental setup is shown in Figure B-1.

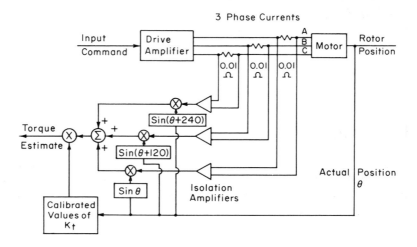

Figure B-1: Experimental setup for the torque estimator

B.2 Estimation and Control Software

```
C
C-----------------------------------------------------------------
C           TORQUE ESTIMATION AND CONTROL
C
C-----------------------------------------------------------------
C
C     IV(1) = volts for PHASE  A  ,
C     IV(2) = volts for PHASE  B  ,
C     IV(3) = volts for PHASE  C  ,
C
      DIMENSION IOSB(2),CURR(3)
      BYTE CHAR(6),TOP(2)
      DATA CHAR/27,91,50,74,27,56/
      DATA TOP/27,56/
      INTEGER  IV(3)
C
C-----------------------------------------------------------------
C     ATD CONVERSION CONSTANTS
C
C     DECIMAL READ IN = 204.8*(VOLTAGE INPUT TO A/D) + 2048.
C     AVG = SLOPE * V0LTS + YINT
C
      SLOPE = 204.8
      YINT  = 2048.0
      IFLAG = 1
      TOREST = 100.0
C
C-------- INITIALIZE CONSTANTS
C
      PI   = 3.141592654
      CONV = 3.141592654/180.0
C     DT = 9.*360./1024. ! INCREMENT IN DEGREE, 10-BIT RES.
      DT = 9.*360./4096. ! INCREMENT IN DEGREE, 12-BIT RES.,
                   ! FOR COMMUTATION
      DTD = 360.0/4096.
      OFF7  = -0.135            ! offset in phase A
      OFF9  = -0.179            ! offset in phase B
      OFF11 = -0.193             ! offset in phase C
C
      ICH07 = 0     ! read phase A current thru chan.  0
      ICH09 = 2     ! read phase B current thru chan.  2
      ICH11 =10      ! read phase C current thru chan. 10
C
```

```
      IUNITO = 0              ! OUPUT PORT control
      IUNITI = 2              ! INPUT PORT position
      MASK  = -1              ! MASK VALUE
C
C---- INQUIRE
C
      WRITE(5,5) CHAR         ! clear screen and reset cursor
  5   FORMAT(1X,4A1)
C
C-------------------------------------------------------------------------
C----------------------- data aquisition ----------------------------
C
 300   continue
      CALL ADINP(IFLAG,ICH07,IVAL07)
      CALL ADINP(IFLAG,ICH09,IVAL09)
      CALL ADINP(IFLAG,ICH11,IVAL11)
      IV(1) = IVAL07          !phase A current = CURR(1)
      IV(2) = IVAL09          !phase B current = CURR(2)
      IV(3) = IVAL11          !phase C current = CURR(3)
C
C   CONVERT TO VOLTS AND CORRECT FOR OFFSETS
C
      DO 101 J = 1,3
      IVAL  = IV(J)
      CURR(J)= ( FLOAT( IVAL ) - YINT ) / SLOPE
 101   CONTINUE
C
      CURR(1) = CURR(1) - OFF7
      CURR(2) = CURR(2) - OFF9
      CURR(3) = CURR(3) - OFF11
C
C---------------------- CALCULATE COMMUTATION CONSTANTS---
C
      CALL DINP(IUNITI,MASK,IOSB,INP)    ! CURRENT POS.
      IPOS = INP.XOR.-1
C
C   POSD  = IPOS*DT
      X    = IPOS*DTD              ! for 9 pairs of poles
      CA   = -SIN( (9.*X  + 60. )*CONV ) ! phase shifts to match
      CB   = -SIN( (9.*X  - 60. )*CONV ) ! generated sine waves
      CC   = -SIN( (9.*X  - 180.)*CONV ) ! to current sine waves
C
C---------------- CALCULATE TORQUE
C
      TOREST = CURR(1)*CA + CURR(2)*CB + CURR(3)*CC
      TOREST = -1.0*TOREST
```

```
      CALL TCONST(CKt,X)          ! Fetch Torque constant
      TOREST = CKt*TOREST         ! estimate torque
C
C-----------------------------------------------------------
C           OUTPUT CONTROL SIGNAL
      CALL CONTROL(TOREST,IERR)
C
      IF ( IERR.GT. 2047 ) IERR =  2047
      IF ( IERR.LE.-2047 ) IERR = -2047
      ICONT = IERR.XOR.-32768    ! Change MSB (8000H)
      CALL DOUT(IUNITO,MASK,IOSB,ICONT)
C
      CALL STOP(ISTOP)
      IF ( ISTOP.EQ.1 )  GO TO 300
      CALL EXIT
      END
C
C
C
```

Appendix C

Single Joint Drive Controller

C.1 Controller

Considering the high noise environment, the controller for a single joints axis consists of an analog velocity loop and a digital position loop. the velocity loop was first optimized based on the actual response of the joint axis. The compensation for the velocity signal of motor 2 of Figure 8-2 is shown in Figure C-1.

The position feedback loop was closed digitally. The 16 bits of actual rotor position are input to the PDP11/60 computer through the DR11K interface. The error between the reference position and the actual position in computed in real time. This setup facilitates the adjustment of the position gain depending on the arm configuration. Furthermore, a saturation algorithm allowed a high gain for the position loop. The controller uses the maximum (minimum) gain when the position error is greater (smaller) than some given limits, and a linear action otherwise. This algorithm was used in controlling the first M.I.T direct-drive arm [Asada and Youcef-Toumi 83a]. With this arrangement, the 16 bit command signal is sent out through the DR11K, however, only a 12 bit Digital to Analog converter is needed. Figure C-2 shows the implementation block.

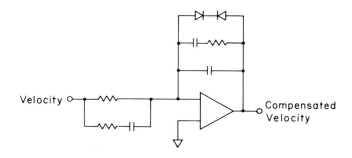

Figure C-1:   Compensation circuit for velocity loop

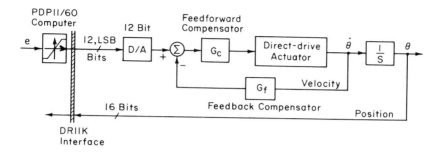

Figure C-2: Controller implementation

C.2 Software

```
C
C-----------------------------------------------------------------------
C          DIRECT-DRIVE ARM CONTROL ROUTINE
C              DIGITAL POSITION LOOP
C-----------------------------------------------------------------------
C
      DIMENSION IOSB(2),IIREF(1000)
      BYTE CHAR(6),TOP(2)
      DATA CHAR/27,91,50,74,27,56/
      DATA TOP/27,56/
      CHARACTER*14 FIN
C
C
      WRITE(5,5) CHAR          ! clear screen and reset cursor
C
C-----------------------------------------------------------------------
C          initialize
C-----------------------------------------------------------------------
C
      TYPE*,'ENTER NUMBER OF LOOPS'
      ACCEPT*,LOOPS
      TYPE*,'ENTER SAMPLING IN SECONDS'
      ACCEPT*,TS
C
      TYPE*,' ENTER 1 FOR DISPLAY '
      ACCEPT*,IDIS
C
C 990    continue
      MASK = -1                     ! MASK VALUE
      IUNITI = 2
      IUNITO = 1
C
C----------------------- READ REFERENCE POSITIONS --------------------
C
      TYPE*,'ENTER FILENAME'
      READ(5,55) FIN
 55    FORMAT(A14)
      OPEN(UNIT=4,STATUS='OLD',NAME=FIN)
      J = 1
 50    READ(4,*,END=501) I,I,IIREF(J)
C      WRITE(5,35) J,J,IIREF(J)
 35    FORMAT(1X,3I12)
      J = J + 1
```

```
         GO TO 50
 501    NDATA = J -1
        CLOSE(UNIT=4)
 C
 C-------------------------------------------------------------------
 C                 set up screen
 C-------------------------------------------------------------------
 C
        WRITE(5,5) CHAR          ! clear screen and reset cursor
 5      FORMAT(1X,4A1)
        WRITE(5,60)
 60     FORMAT(/,10X,
     1 '   <---REQUEST POSITION: integer   ',/,10X,
     1 '   <---POSITION ERROR  : integer   ',/,10X,
     1 '   <---ACTUAL  POSITION: integer   ',/,10X,
     1 '   <---counter                ',/,10X,
     1 '   <---time                 ',/,10X,
     1 '     PRESS ANY KEY FOR NEXT POSITION')
 c      WRITE(5,62) IREF,IERR,IPOS
 c
 62     FORMAT(/,1X,I10,/,1X,I10,/,1X,I10,/,1X,I10,
     1      /,1X,F10.3,/,1x)
 63     FORMAT(1X,2A1)
 C
 C-------------------------------------------------------------------
 C             READ AND WRITE POSITIONS
 C             POSITION LOOP
 C-------------------------------------------------------------------
 C
        type*,'PRESS RETURN TO START CONTROLLER'
        READ(5,51)
 51     FORMAT(I1)
 C
        JJ = 0
 C
 777    JJ = JJ + 1
 C
 C
        j = 1
 666        IREF = IIREF(J)   ! SET HOME POSITION = 3737
 C
            TI = SECNDS(0.0)
 C
 400             CONTINUE
                CALL DINP(IUNITI,MASK,IOSB,IPOS)
                IERR  =  IPOS - IREF
```

```
          CALL GAIN(IPOS,KP)
          ICONT = IERR*KP
          IF ( IERR.GT. 2047 ) IERR = 2047
          IF ( IERR.LE.-2047 ) IERR = -2047
          ICONT = IERR.XOR.-32768 ! change MSB (8000H)
          CALL DOUT(IUNITO,MASK,IOSB,ICONT)
      IF (IDIS.EQ.1 ) THEN
      WRITE(5,63) TOP              ! reset cursor to home
      WRITE(5,62) IREF,IERR,IPOS,J,TF
      END IF
C
C
      TF = SECNDS(0.0) - TI
      IF ( TF.LE.TS ) GO TO 400
      IF ( J.EQ.NDATA.and.JJ.EQ.LOOPS ) GO TO 400
      IF ( J.EQ.NDATA.and.JJ.NE.LOOPS ) GO TO 777
      J = J + 1
    GO TO 666
C
C-----------------------------------------------------------------
C                   exit
C-----------------------------------------------------------------
C
C 991   TYPE*,'YOU HAVE PROBLEMS'
C    TYPE*,'STATUS CODE:    ',IOSB(1),IOSB(2)
C
 999   CALL EXIT
      END
C
C
```

Appendix D

The Jacobian Matrix

The end-effector and joint motions can be represented by two vectors $\mathbf{x}$ and θ respectively. The kinematic equations of a manipulator relate the end-effector positions and orientations to the manipulator joint displacements:

$$\mathbf{x} = \mathbf{f}(\theta) \qquad\qquad (D.1)$$

These kinematic equations can be solved to determine the joint displacements θ_o that correspond to an end-effector position and orienation $\mathbf{x}_o$. In general, we are interested in not only the instantaneous position and orientation of the end-effector but also its velocity. For this, we need to derive the differential relationship that exists between the first joint and end-effector displacements. This differential relationship is obtained by differentiating Equ. (D.1):

$$dx_1 = \frac{\partial f_1}{\partial \theta_1} d\theta_1 + \frac{\partial f_1}{\partial \theta_2} d\theta_2 + \cdots + \frac{\partial f_1}{\partial \theta_n} d\theta_n$$

.
.
.

$$dx_m = \frac{\partial f_m}{\partial \theta_1} d\theta_1 + \frac{\partial f_m}{\partial \theta_2} d\theta_2 + \cdots + \frac{\partial f_m}{\partial \theta_n} d\theta_n$$

where $dx^T = [dx_1, \ldots, dx_m]$, $d\theta^T = [d\theta_1, \ldots, d\theta_n]$ are the differential displacements and $[f_1, \ldots, f_n]$ are the components of the vector function $\mathbf{f}$.

The partial derivatives of the functions $f_1(\theta), \ldots, f_m(\theta)$ can be organized in a matrix $\mathbf{J}$ given by:

$$
J = \begin{bmatrix} \dfrac{\partial f_1}{\partial \theta_1} & \cdots & \dfrac{\partial f_1}{\partial \theta_n} \\ \\ \dfrac{\partial f_m}{\partial \theta_1} & \cdots & \dfrac{\partial f_m}{\partial \theta_n} \end{bmatrix}
$$

This $m \times n$ matrix, J, is referred to as the manipulator Jacobian matrix. Thus the differential displacements dx and $d\theta$ are related by J:

$$
dx = J d\theta \tag{D.2}
$$

Equ. (D.2) can be divided through by the time differential dt to arrive at:

$$
\frac{dx}{dt} = J \frac{d\theta}{dt}
$$

Therefore, the Jacobian matrix also relates the joint velocities to the end-effector velocities.

References

[Asada 83]   Asada H.
A Geometrical Representation of Manipulator Dynamics and
    its Application to Arm Design.
*ASME Journal of dynamic Systems, Measurement, and
    Control* 104-3, 1983.

[Asada and Slotine 86]
         Asada, H. and Slotine, J-J.E.
Robot Analysis and Control.
John Wiley & Sons, 1986.

[Asada and Kanade 83]
         Asada, H., and Kanade, T.
Design of Direct-Drive Mechanical Arms.
*ASME Journal of Vibration, Acoustics, Stress, and reliability
    in design* 105(3):312-316, 1983.

[Asada and Lim 85]
         Asada, H. and Lim, S.
Design of Joint Torque Sensor and Torque Feedback Control
    for Direct-Drive Arms.
In *the Proc. of the ASME, Winter Annual Meeting.* Miami,
    Florida, Nov., 1985.

[Asada and Ro 85]
         Asada, H. and Ro, I.H.
A Linkage Design for Direct-Drive Arms.
*ASME Journal of Mechanisms, Transmission and
    Automation in Design* 107, 1985.

[Asada and Youcef-Toumi 83a]
         Asada, H., and Youcef-Toumi, K.
Analysis and Design of Semi-Direct-Drive Robot Arms.
In *Proc. of 1983 American Control Conference*, pages
    757-764. San Francisco, 1983.

[Asada and Youcef-Toumi 83b]
         Asada, H., and Youcef-Toumi, K.
Analysis of Multi-Degree-of-Freedom Actuator Systems For
    Robot Arm Design.
In *Prod. of 1983 ASME Winter Annual Meeting.* 1983.

[Asada and Youcef-Toumi 84]
                Asada, H., and Youcef-Toumi, K.
                Analysis and Design of a Direct-Drive Arm With a Five-Bar-
                    Link Parallel Drive Mechanism.
                *ASME Journal of Dynamic Systems, Measurement and
                    Control* 106(3):225-230, 1984.

[Asada and Youcef-Toumi 84]
                Asada, H. and Youcef-Toumi, K.
                Decoupling of Manipulator Inertia Tensor by Mass
                    Redistribution.
                In *Proceeding of the ASME Mechanisms Conference.*
                    October, 1984.

[Asada, Kanade and Takeyama 83]
                Asada, H., Kanade, T., and Takeyama, I.
                Control of a Direct-Drive Arm .
                *ASME Journal of Dynamic Systems, Measurement, and
                    Control* 105(3):136-142, 1983.

[Asada, Youcef-Toumi and Lim 84]
                Asada, H., Youcef-Toumi, K. and Lim, S.
                Joint Torque Measurement of a Direct-Drive Arm.
                In *Proc. of the 23rd Conference on Decision and Control.*
                    Las Vegas, NV., Dec., 1984.

[Asada, Youcef-Toumi and Ramirez 84]
                Asada, H., Youcef-Toumi, K. and Ramirez, R.
                Design of M.I.T Direct-Drive Arm.
                In *International Symposium on Design and Synthesis, Japan
                    Society of Precision Engineering.* Tokyo, Japan, July,
                    1984.

[Atkeson, An, and Hollerbach 85]
                Atkeson, C.G., An, C.H., and Hollerbach, J.M.
                Estimation of Inertial Parameters of Manipulator Loads and
                    Links.
                In *Prod. of Third International Symposium of Robotics
                    Research*, pages 32-39. 1985.

[Davis and Chen 84]
                Davis, S. and Chen, D.
                High Performance Brushless DC Motors for Direct-Drive
                    Robot Arms.
                In *Proceedings of IEEE.* , Oct., 1984.

[Denavit and Hartenberg 55]
> Denavit, J. and Hartenberg, R. S.
> A Kinematic Notation for Lower-Pair Mechanisms Based on
> Matrices.
> *ASME Journal of Applied Mechanics* :215-221, 1955.

[Dubowsky and Des Forges 79]
> Dubowsky, S., and Des Forges, D. T.
> The Application of Model-Referenced Adaptive Control to
> Robotic Manipulators.
> *ASME Journal of Dynamic Systems, Measurement and
> Control* 101:193-200, 1979.

[Electro-Craft 80] Electro-Craft.
> *DC Motors,Speed Controls,Servo-Systems* .
> Engineering Handbook by Electro-Craft Corporation, Minn.,
> 1980.

[Freund 82]   Freund, F.
> Fast Nonlinear Control with Arbitrary Pole Placement for
> Industrial Robots and Manipulators.
> *Int. Journ. of Robotics Research* 1(1):65-78, 1982.

[Hanafusa and Asada 77]
> Hanafusa, H., Asada, H.
> A Robot Hand with Elastic Fingers and Its Application to
> Assembly Process.
> In *Proc. of IFAC First Symposium on Information Control
> Problems in Manufacturing Technology*, pages 127-138.
> Tokyo, 1977.

[Hogan 85]   Hogan, N.
> Impedance Control.
> *ASME Journal of Dynamic Systems, Measurement and
> Control* 107(1), March, 1985.

[Hollerbach 80]   Hollerbach, J. M. .
> A Recursive Lagrangian Formulation of Manipulator
> Dynamics and a Comparative Study of Dynamics
> Formulation complexity.
> In *IEEE Trans. on Systems, Man, and Cybernetics*, pages
> 730-736. November, 1980.

[Horn and Raibert 78]
> Horn, B. H. K., and Raibert, M. H.
> Configuration Space Control.
> In *The Industrial Robot*, pages 66-73. 1978.

[Horowitz and Tomizuka 81]
Horowitz, R., and Tomizuka, M.
An Adaptive Control Scheme for Mechanical Manipulators -
Compensation of Nonlinearity and Decoupling Control.
*ASME Journal of dynamic Systems, Measurement, and
Control* , 1981.

[Inoue 71]   Inoue, H.
Computer Controlled Bilateral Manipulator .
In *Bulletin*, pages 199-207. Japan Soc. Mech. Eng., 1971.

[Kanade, Khosla and Tanaka 84]
Kanade, T., Khosla, P. K. and Tanaka, N.
Real-Time Control of CMU Direct-Drive Arm II Using
Customized Inverse Dynamics.
In 23rd *Conference on Decision and Control.* IEEE,
December, 1984.

[Kane 68]   Kane, T. R.
Dynamics.
Holt, Rinehart and Winston, Inc., 1968.

[Leborgne, Ibarra, and Espiau 81]
LeBorgne, M., Ibarra, J. M., and Expiau, B.
Adaptive Control of High Velocity Manipulators.
In *Proc. of the 11-th Int. Symp. on Industrial Robots*, pages
227-236. Tokyo, Oct., 1981.

[Luh, Fisher, and Paul 83]
Luh, J. Y. S., Fisher, W. D. and Paul, R. P.
Joint Torque Control by a Direct Feedback for Industrial
Robots.
IEEE Transactions on Automatic Control AC-28:153-161,
Feburary, 1983.

[Luh, Walker, and Paul 80]
Luh, J. Y. S., Walker, M. W., Paul, R. P. C.
On Line Computational Scheme for Mechanical
Manipulators.
*ASME Journal of Dynamic Systems, Measurement and
Control* 102, 1980.

[Mason 81]   Mason, M.
Compliance and Force Control for Computer Controlled
Manipulators.
IEEE Transactions on Systems, Man, and Cybernetics
SMC-11(6), June, 1981.

[Mayeda, Osuka, and Kangawa 84]
Mayeda H., Osuka K., and Kangawa A.
A New Identification Method for Special Manipulator Arms.
In *Prod. of 9th World Congress of the International Federation of Automatic Control*, pages 74-79. 1984.

[Mukerjee and Ballard 85]
Mukerjee,A. and Ballard, D.H.
Self-Calibration in Robot Manipulators.
In *IEEE International Conference on Robotics and Automation*, pages 1050-1057. March, 1985.

[Paul 81]
Paul, R.
Robot Manipulators: mathematics, Programming, and Control.
M.I.T Press, 1981.

[Paul and Shimano 76]
Paul, R. and Shimano, B.
Compliance and Control.
In *Proc Joint Automat Contr., San Francisco*, pages 694-699. 1976.

[Paynter 82]
Paynter, H. M.
Modeling and Simulation of Dynamic Systems.
In *Course 2.141, Department of Mechanical Engineering, Massachusetts Institute of Technology.* Spring, 1982.

[Raibert and Craig 81]
Raibert, M.H. and Craig, J.J.
Hybrid Position/Force Control of Manipulators.
Trans ASME J. Dynamic Syst., Meas., Contr. 103:126-133, June, 1981.

[Ramirez 84]
Ramirez, R. .
Design of High Speed Graphite Composite Robot Arm.
In *Master of Science Thesis, Department of Mechanical Engineering.* Massachusetts Institute of Technology, Feb., 1984.

[Salisbury 80]
Salisbury, J. K.
Active Stiffness Control of a Manipulator in Cartesian Coordinates.
In *19th IEEE Conf on Decision Control*, pages 95-100. December, 1980.

[Slotine and Sastry 83]
              Slotine and Sastry.
              Tracking control of Non-linear Systems Using Sliding
                   Surfaces, with Applications to Robot Manipulators.
              Int. Journal of Control 38-2:465-492, 1983.

[Stepanenko and Vukobratovic 76]
              Stepanenko, Y., and Vukobratovic, K.M.
              Dynamics of Articulated Open-Chain Active Mechanisms.
              In *Math. Biosci.*, pages 137-170. 1976.

[Strang 80]   Strang, G.
              Linear Algebra and its Applications.
              Academic Pres, Inc., 1980.

[Takase 77]   Takase, K.
              Task-Oriented Variable Control of Manipulator and its
                   Software Servoing System.
              In *IFAC Symp Information and Control Problems in
                   Manufacturing Technology, Tokyo.* 1977.

[Takegaki and Arimoto 81]
              Takegaki, M. and Arimoto,S.
              An Adaptive Trajectory Control of Manipulators.
              Int. Journal of Control 34(2):219-230, 1981.

[Tal 81]   Tal, J.
              Modeling Motors for Control Applications.
              In *Proceedings of the First Int. MOTORCON Conf.*, pages
                   2A.1-1 - 2A.1-12. , June, 1981.

[Tani, Hatamura, and Nagao 83]
              Tani, Y., Hatamura,Y. and Nagao, T.
              Developement of Small Three Component Dynamometer for
                   Cutting Force Measurement.
              Bulletin of the JSME 26(214), April, 1983.

[Tepper and Lowen 72]
              Tepper, F. R. and Lowen, G. G.
              General Theorems Concerning Full Force Balancing of
                   Planar Linkages by Mass Redistribution.
              Journal of Eng. Ind. , Trans. of ASME Ser., B,94:789-796,
                   1972.

[Tepper and Lowen 73]
        Tepper, F. R. and Lowen, G. G.
        Two General Rules for Full Force Balancing of Planar
           Linkages.
        In *Proc. 3rd. Appl. Mech. Conf., Paper No 10..* Oklahoma
           State University, 1973.

[Tepper and Lowen 75]
        Tepper, F. R. and Lowen, G. G.
        Shaking Force Optimization of Four-Bar Linkages with
           Adjustable Constraints on Ground Bearing Forces.
        Journal of Eng. Ind. , Trans. of ASME Ser., B,97:643-651,
           1975.

[Utkin 72]   Utkin, V.I.
        Equations of Sliding Mode in Discontinuous Systems.
        Automation and Remote Control I(II), 1972.

[Utkin 77]   Utkin, V.I.
        Variable Structure Systems with Sliding Mode: A Survey.
        IEEE Transactions on Automatic Control AC-22, 1977.

[Utkin 78]   Utkin, V.I.
        Sliding Modes and Their Application to Variable Structure
           Systems.
        MIR Publishers, Moscow, 1978.

[Vukobratovic and Stokic 80]
        Vukobratovic, K.M., and Stokic, D.M.
        Contribution to the Decoupled Control of Large-Scale
           Mechanical Systems.
        In *Automatica*, pages 9-21. January, 1980.

[Vukobratovic and Stokic 81]
        Vukobratovic, K.M., and Stokic, D.M.
        One Engineering Concept of Dynamic Control of
           Manipulators.
        ASME Journal of Dynamic Systems, Measurement, and
           Control 102:108-118, 1981.

[Walker and Oldham 78]
        Walker, M. J. and Oldham, K.
        Mechanisms and Machine Theory.
        1978, pages 175-185.

[Weichbrodt and Beckman 78]
              Weichbrodt, B. and Beckman, L.
              Some Special Applications for ASEA Robots - Deburring of
                   Metal Parts in Production.
              In *Robots III*, pages 349-361. 1978.

[Welburn 84]   Welburn, R.
                 Ultra High Torque Motor System for Direct Drive Robotics.
                 In *Robot 8, Detroit*, pages 19-63 - 19-71. June, 1984.

[Whitney 77]   Whitney, D.E.
                 Force Feedback Control of Manipulator Fine Motions.
                 *Trans. ASME Journal of Dynamic Systems, Measurement,
                     and Control* 99:91-97, 1977.

[Wu and Paul 80] Wu, C. H. and Paul, R. P. .
                 Manupulator Compliance based on Joint Torque Control.
                 In *19th IEEE Conf on Decision Control*, pages 88-94.
                 December, 1980.

[Yang and Tzeng 85]
              Yang, D.C.H and Tzeng, S.W.
              Simplification and Linearization of Manipulator Dynamics
                   by Design.
              In *Proc. of the 9th Applied Mechanics Conference.* October,
                   1985.

[Yang and Tzeng 86]
              Yang, D.C.H and Tzeng, S.W.
              Simplification and Linearization of Manipulator Dynamics
                   by The Design of Inertia Distributions.
              Int. Journ. of Robotics Research 5(3): , 1986.

[Youcef-Toumi 85]
              Youcef-Toumi, K.
              Analysis, Design and Control of Direct-Drive Manipulators.
              In *Doctor of Science Thesis, Department of Mechanical
                   Engineering.* Massachusetts Institute of Technology,
                   May, 1985.

[Youcef-Toumi and Asada 85a]
              Youcef-Toumi, K. and Asada, H.
              The Design of Arm Linkages With Decoupled and
                   Configuration-Invariant Inertia Tensors Part I: Open
                   Kinematic Chains With Serial Drive mechanisms.
              In *Proc. of 1985 ASME, Winter Annual Meeting, Miami,
                   Florida.* November, 1985.

[Youcef-Toumi and Asada 85b]
Youcef-Toumi, K. and Asada, H.
The Design of Arm Linkages With Decoupled and
Configuration-Invariant Inertia Tensors Part II: Actuator
Relocation and Mass Distribution.
In *Proc. of 1985 ASME, Winter Annual Meeting, Miami,
Florida*. November, 1985.

[Youcef-Toumi and Asada 86]
Youcef-Toumi, K. and Asada, H.
The Design of Open-loop Manipulator Arms With
Decoupled and Configuration-invariant Inertia Tensors.
In *Proceeding of the 1986 IEEE International Conference on
Robotics and Automation*. April, 1986.

[Youcef-Toumi and Asada 85]
Youcef-Toumi, K. and Asada, H. .
Dynamic Decoupling and Control of a Direct-Drive
Manipulator.
In *Proceeding of IEEE conference on Decision and Control*.
December, 1985.

[Young 78]   Young, K.-K.-D.
Controller Design for a Manipulator Using the Theory of
Variable Structure Systems.
IEEE Transactions on Systems, Man and Cybernetics 8(2),
1978.

Index